Felix Publishing 2018
www.felixpublishing.com.au
email: info@felixpublishing.com
Print copies available from publisher.

I0134904

Through Sea and Sky
Part of the Series **Adventures in Earth Science**
Other books in the series include:
> Exploration Science (Field Geology and Mapping)
> Riches from the Earth (Earth Materials and Mining)
> Changing the Surface (Erosion and Landscapes)
> Rocks – Building the Earth
> Fossils – Life in the Rocks
> A Dangerous Planet (Earth Hazards)
> Beyond Planet Earth (Astronomy)

2018 digital book release
ISBN: 978-0-9946432-1-6
Second Print Edition
ISBN: 978-0-9946432-2-3

Author: Dr. Peter T. Scott
All illustrations, photographs and videos by the Author unless stated
Cover photo: Cloud formation over Port Denaru, Fiji. Design suggested by AJS Creative.

Registration:
Thorpe-Bowker +61 3 8517 8342
email: bowkerlink@thorpe.com.au

FELIX

To my grandchildren who are
yet to find their own adventures.

THROUGH

SEA and SKY

Dr. Peter T. Scott

First released 2017
All rights reserved Felix Publishing

About the Author

Dr. Scott as a crewman on the bowsprit of the brigantine *One-and-All*, somewhere in the Coral Sea

Dr. Peter Scott is an award-winning teacher of Earth Science of over forty years' experience in both Secondary and Tertiary Education. He holds a Bachelors' Degree, two Masters' Degrees and a Doctorate including several years study in oceanography and meteorology. A keen sailor, he has sailed skiffs around Sydney Harbour and Botany Bay in Australia and for many years also cruised in his own small yacht out of Brisbane. With a love for the sea, he had several cruises in schooners and brigantines along most of the Australian eastern coast, especially in the Great Barrier Reef. He was an Officer/Instructor in the Australian Naval Cadets for over twelve years including as Captain of Training Ship *Magnus*.

Table of Contents

CHAPTER 1: EXPLORING THE SEAS

1.1 Introduction

Oceanography is the study of the sea and the sea-floor, that is, it embodies all studies concerned with the chemical and physical nature of seawater, ocean currents and seafloor **topography** and structure. There is considerable overlapping with **hydrography** which is more an applied science dealing with the waters of the Earth, their mapping and study for trade and economic use.

Oceanography is useful in explaining the formation of:

- Marine sediments and fossils found in strata on land
- Sea-floor topography – the structures found on the sea-floor
- Ocean currents and sea temperature and
- Crustal plates and their movement.

In addition, Oceanography has provided a better understanding of:

- Volcanism and earthquakes
- Environmental changes
- Ecological relationships within the oceans and
- Natural resources such as minerals, oil and gas.

1.2 The Tides

Tides are the apparent changes in the local levels of the sea and are due to the gravitational pull of the Moon on the waters of the Earth and the planet's rotation.

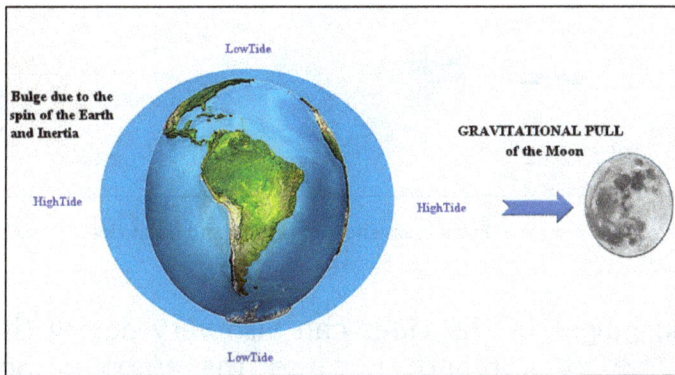

Figure 1.1: Diagram showing how the Moon's gravitational pull produces two high and low tides on opposite sides of the Earth

Even though the Sun is a very long way from the Earth (149.6 million km.) compared to the Moon (384,400 km), it does exert a small gravitational force on the oceans. This causes slight variations in the tides which are noticeable when the Sun and the Full Moon are in the same line, giving high tides slightly higher than usual called **spring tides** – as the water springs up, not because of the season. When the Sun and Quarter Moon are at right angles, high tides are lower than is usual and are called **neap tides** from the Old English _nēp_ meaning scant or low.

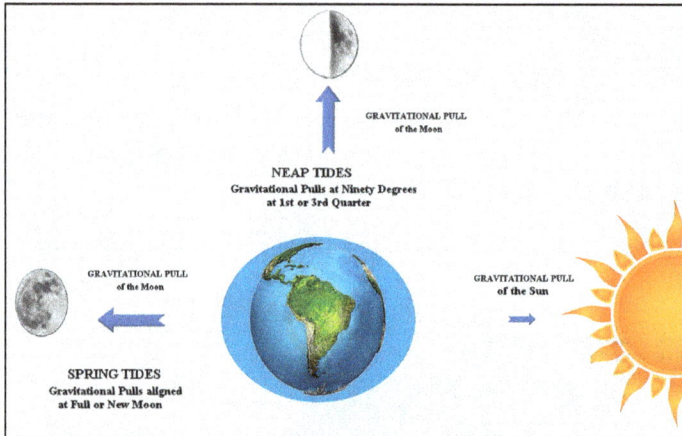

Figure 1.2: Diagram showing spring and neap tides

The height of the tides can also vary during the course of a month because the Moon is not always the same distance from the Earth. As the Moon's orbit brings it in closer proximity to our planet (its **perigee**), its gravitational forces can increase by almost 50%, and this stronger force leads to higher tides. Likewise, when the Moon is farther away from the Earth (its **apogee**), the tides are not as high.

Tides most usually occur twice a day as one high tide and one low tide as **diurnal tides. Semi-**

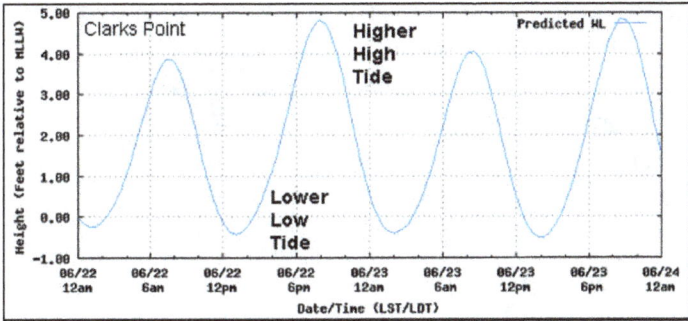

Figure 1.3: A tidal chart for Clark's Point, New Bedford, USA (NOAA)

diurnal tides can also occur, giving two high waters and two low waters each day. These highs and lows do not happen at the same time each day because the Moon takes slightly longer than 24 hours to line up again exactly with the same point on the Earth (about 50 minutes more). Thus the timings of highs and lows are staggered throughout the course of a month, with each tide commencing approximately 24 hours and 50 minutes later than the one before it.

1.3 Navigation - the First Skills

Oceanography developed from the need for mariners to know more about of the sea and what it contains and started with their practices and voyages. Navigation was perhaps the first skill which developed along scientific lines of observation, innovation and practice. In the Pacific, Micronesian and Polynesian navigators crossed long distances using the stars, climatic conditions and the direction, size and reflected patterns of wave swells to move from one island to another. Such information was recorded using coconut fibre memory aids often wrongly called stick charts.

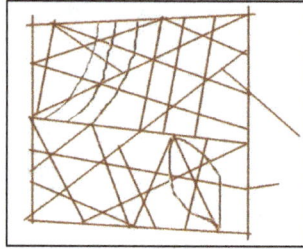

Figure 1.4: A reconstruction of a Pacific Islander stick chart

In Europe and Asia, ships rarely strayed far from the coast and so a detailed knowledge of these coastal trading routes became known through word-of-mouth and later in early writings. To aid them in finding a general direction, the early mariners often used the stars. They had noted that whilst most stars rose in the East and travelled across the sky following a regular path and time, one star, Polaris or the North Star, appeared to stay in the one place in the sky. This was because this star is directly over the North axis of the Earth's rotation. Once found, it was a useful guide in following a course at night.

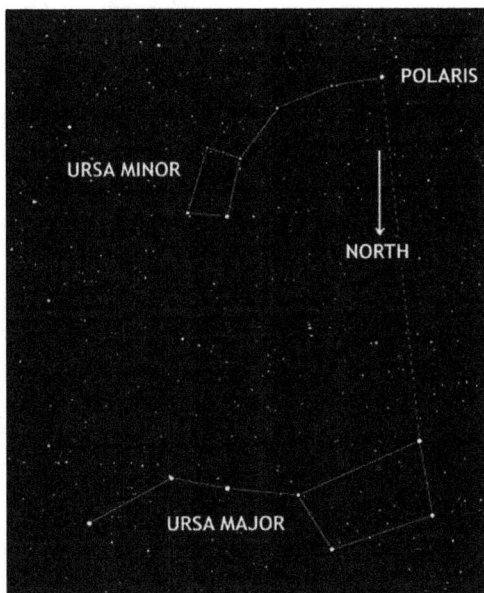

Figure 1.5: Finding Polaris in the Northern Hemisphere - subtend a line from the "Big Dipper" (the Constellation of the Big Bear - Ursa Major) to the end of the "Little Dipper" (Ursa Minor).

In the Southern Hemisphere, early European explorers found no such guiding star. Eventually they made use of the Southern Cross (the constellation Crux – the Cross) to locate South. To find south using the Southern Cross in the Southern Hemisphere, subtend a line between the two pointers (Alpha and Beta Centaurii) and one from the long axis of the Cross. Where they meet is the **South Celestial Pole (SC.P)**, the imaginary spot immediately above south which is on the horizon below.

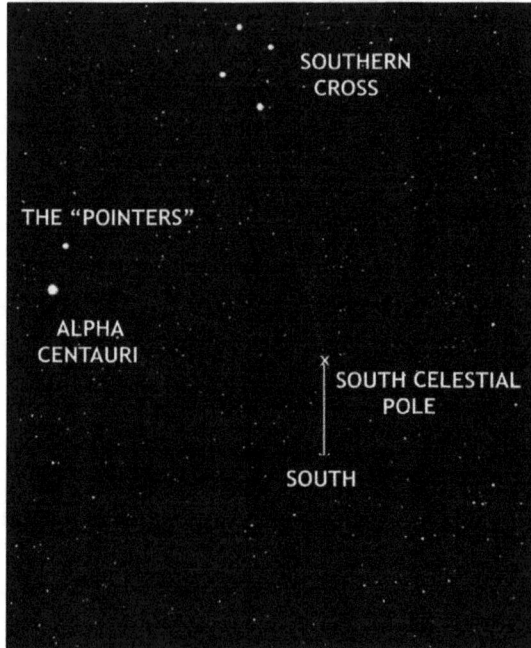

Figure 1.6: Finding South with the Southern Cross is done by taking a line from the central position between the two pointers and extending it to where it meets a line taken through the long arm of the cross. Drop a line straight down from this South Celestrial Pole to south on the horizon

Alternatively, imagine a line drawn through the long axis of the Cross and go four and one half times the length of the axis to the South Celestial Pole. This is a quick method to check one's bearing generally even as the Cross rotates around the S.C.P. However, this does not work well in summer when the Cross is very low on the horizon and often obscured by trees, mountains and clouds.

Later, as mariners ventured further away from the visible coast, navigational instruments and aids were developed to give a more accurate position than the general position of the Sun or stars although these reference points are still good for a quick and rough orientation check. The Norse (Vikings) used a simple sun compass with a vertical **gnomon** and a notched circle marked with course positions at times of day. This enabled the Norse to travel westward from Norway and the Baltic to Iceland, Greenland and the mainland of North America (Vinland). When the Sun was obscured by cloud, fog or if it had just set, the Norse Iceland spar may have used their Sun Compass with two clear crystals of Calcite or, which they called a **sólarsteinn** (Icelandic for sunstone), which would show the position of the Sun by the polarizing effect of that crystals even after sunset around the spring equinox (in March in the Northern hemisphere when the times of day and night are approximately equal).

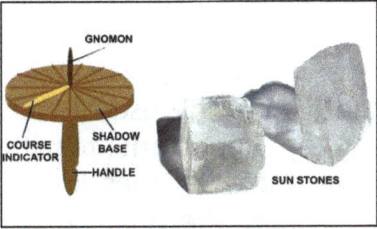

Figure 1.7: Sun compass (left) and sunstones (right)

GNOMON

COURSE INDICATOR

SHADOW BASE

HANDLE

SUN STONES

Outward bound, the Norse usually travelled west, keeping the Sun generally to their left. This was on the **starboard** going West– the side where the steering oar was located; from the Old Norse stýri for rudder, and borð for the side of a ship. The name **larboard,** meaning the loading side well away from the steering oar, on the other side of the ship was changed to **port** to prevent confusion. The use of the Sun's position was thus a very important maritime skill and travelers from ancient times knew that at any given date and time, the angle of the Sun changed as they moved South or North. **Eratosthenes of Cyrene** (Greek: 276–194 BC) was the chief librarian at the Library of Alexandria in Egypt, who used this knowledge to calculate a value for the Earth's circumference; the Greeks already assumed that the Earth was a sphere based on its shadow on the Moon during Lunar Eclipses). He also measured the tilt of the Earth's axis or its **obliquity**, at around 23 degrees from the vertical and used navigational lines on maps.

Early mariners soon used the angle of the Sun at noon when it was at its highest point, to find how far south or north they had travelled. Instruments were soon developed to measure the angle of the Sun above the horizon and to take sightings from stars and other nearby objects such as coastal features. The simple Jacob's Staff was replaced in the Middle Ages by the mariners' astrolabe. This was the forerunner of the octant and later the more efficient sextant.

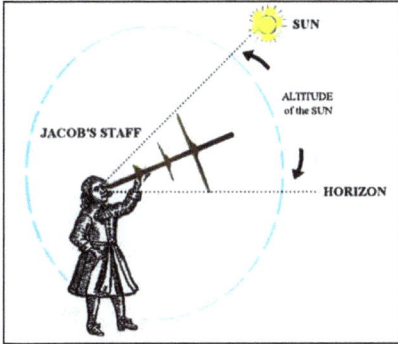

Figure 1.8: An early Jacob's Staff which was used from the time of the Ancient Greeks to measure angles of stars to the horizontal

Figure 1.9: A 16th Century mariner's astrolabe – it is held vertical by the top ring and sighted along the central rotating arm.

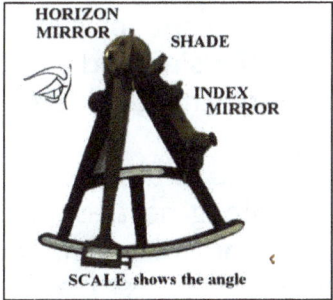

Figures 1.10 & 1.11: The octant (left) was developed independently by John Hadley (1682-1744) and Thomas Godfrey (1704-1749) from Isaac Newton's (1643-1727) Reflecting Quadrant of 1699. These were later refined in the 19th Century into the sextant (right), a more accurate and compact device with a telescope for good sighting, Sun filters and better mirrors.

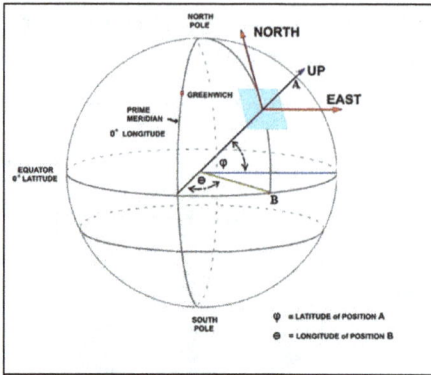

Figure 1.12: Diagram showing Latitude and Longitude

These devices were useful in finding the **latitude**, or distances north or south of the Equator, along east-west lines circling the Earth known as parallels (to the **Equator**) by measuring the angle of the Sun above the horizon at noon. They could also be used to find the latitude from stars using sets of complex tables which gave the times of rising and positions of stars at different places on the Earth's surface.

Longitude is the position east or west around the Earth along north-south circles around the globe called **meridians.** In early seafaring, longitude could only be estimated using the rotation of the Earth and some device for measuring time. Mariners knew that along a certain line of latitude, they would reach a destination or need to turn on another bearing after so many days, as each hour equals 15 degrees longitude of Earth rotation. This method of estimation or **dead reckoning** was often not accurate enough. For example, the Dutch East India Company traders of the 17th Century would leave their colony at Cape Town, South Africa, and sail for a given number of days eastward with the Roaring Forties winds along 40 degrees of latitude then swing north and sail for their colony of Java, now in Indonesia. If their timing was wrong, they would continue sailing east until wrecked on the rugged coast of Western Australia which is now littered with such wrecks.

Latitude and longitude, being angles subtended from the centre of the Earth, are measured in degrees (^{0}), minutes (') and seconds (") of arc. There are 60 seconds of arc (or arcseconds - not to be confused with seconds of time) in one minute of arc and 60 minutes in one degree. Latitude is measured from the Equator which is at zero degrees, to each of the geographical poles at 90 degrees north or south of the Equator. Longitude is measured from the **Prime Meridian** at Greenwich near London at zero degrees, to east or west until 180 degrees on the other side of the globe at the **International Date Line (IDL).** These coordinates may also be given in degrees and decimal fractions of a degree, For example, the coordinates for some sea ports are:

PORT	LATITUDE Always given first	LONGITUDE
Sydney (Australia)	33.8688° South	151.2093° East
Brisbane (Australia)	27.4698° South	153.0251° East.
Valparaiso (Chile)	33.0472° South	71.6127° West
Rio de Janeiro (Brazil)	22.9068° South	43.1729° West
Cape Town (South Africa)	33.9249° South	18.4241° East
New York (USA)	40.7128° North	74.0059° West
San Francisco (USA)	37.7749° North	122.4194° West
London (united Kingdom)	51.5074° North	0.1278° West
Liverpool (United Kingdom)	53.4084° North	2.9916° West
Singapore	1.3521° North	103.8198° East
Tokyo (Japan)	35.6895° North	139.6917° East
Shanghai (China)	31.2304° North	121.4737° East
Hong Kong (China)	22.3964° North	114.1095° East

Table 1.1 Geographical coordinates for some ports

Early clocks taken to sea were very inaccurate, being subjected to the continual roll of the ship and saltwater. It was not until the 1761 H4 **chronometer,** from the Greek Chronos for time and meter to measure, of **John Harrison** (English: 1693 - 1776) that these problems were overcome. Then an accurate measurement of Longitude could be made based on the Earth's regular 24 hour rotation with 1 hour of time equals 15 degrees of Longitude. In 1851, the

Royal Observatory at Greenwich near London was established as the prime Meridian as zero degrees longitude from which all other distances could be taken.

Figure 1.13: A 19th century ship's chronometer in its protective case and mounted on gimbals

1.4 The Magnetic Compass

The Chinese had found that a naturally-occurring mineral, magnetite, also called lodestone, would always swing to the same direction if suspended by a cord, balanced on a board or floated on a small wooden raft in a bowl of water. They used this first magnetic compass to find south and gradually this innovation became known in Europe via the Middle around the 12th Century. Over time, the compass improved and became an essential part of navigation at sea as well as on land and in the air.

Figure 1.14: Replica of a Han Dynasty (206 BC- 220 AD) Spoon Compass. The finely-balanced spoon was made from lodestone and is constructed so that the handle points south as its cardinal (or reference) direction. (Photo: Wiki Commons)

It soon became apparent that there were some errors associated with the magnetic compass. For one thing, Magnetic North did not exactly match Geographic North. Whilst the Geographical North Pole is at the axis of tilt of the Earth and the place where cartographers have converged all of the meridians of Longitude, the **Magnetic North Pole** is where the Earth's magnetism is the most strongest and where, if held freely, a compass needle would point down towards the ground, hence such devices are called magnetic dip compasses. There are also corresponding South Magnetic and South Geographic Poles.

The Earth's magnetic field is believed to be caused by part of the Earth's core being molten nickel-iron. This hot fluid is in motion and is highly electrically charged which in turn produces a magnetic field. Instability in the core and the motion of the Earth means that the magnetic poles are not fixed. Once located near Ellesmere Island in the far north of Canada, the North Magnetic Pole has moved further towards the North geographic Pole and Siberia at about 55 km. per year (data from **NOAA**) – an effect known as **polar wandering.**

Geological evidence from magnetized haematite crystals (a magnetic iron oxide) in basalt rock also shows variations in position for the poles over time, and there have been indications that there may have been several complete reversals in geomagnetism over geological time. Also the direction of the magnetic field lines between magnetic north and south are not uniform across the surface of the Earth and adjustments must be made so that navigators can locate their geographical positions in respect to this magnetic field. With such variation in magnetic compass reading, mariners needed to add or subtract the **magnetic declination**, also called magnetic variation, to their to their bearings at their location with respect to the geographical coordinates to compensate for this difference between the North Magnetic Pole and the North Geographic Pole. The values of this declination will change year by year so it is important to look at the declination diagram at the bottom of the navigation chart to calculate the current value for that locality. Many GPS and other navigations systems are able to be adjusted for True or Geographical North and there are internet pages which will show local variations. Navigation at sea is complex and beyond the scope of this book, but using charts and compasses at sea involves many of the basic principles of land navigation outlined in the companion book EXPLORATION SCIENCE.

Another error of the magnetic compass is the attraction of its needle to any nearby source of Iron – either the metal or some large Iron mineral deposit such as magnetite or haematite. This is called **magnetic deviation** and it became more of a problem as ships incorporated more and more Iron and steel into their design. To use the compass with some degree of accuracy, compensators, iron spheres, were placed on each side of the binnacle which held the compass. These could be moved closer to each side as desired to obtain an accurate North reading when the ship was physically aligned with selected bearings using an isolated onshore compass and reference points. This was called "swinging the ship".

Figure 1.16: A compass binnacle and its red (port side) and green (starboard) compensators. These are moved in or out to reset the compass needle against the magnetic influences of the ship so that the compass needle will point north

This problem became more acute with the advent of modern iron-hulled ships which also had onboard many electrical circuits generating their own magnetic fields. To overcome this problem, the **gyrocompass** was developed at the beginning of the 20th Century. These use a gyroscope, a heavily-weighted device mounted on gimbals which can allow rotation in three dimensions. When the gyroscope is rapidly spun using an electric motor, it will resist any attempt to change its position due to its **inertia** and respond to the Earth's rotation by its **precession**. Gyrocompasses can be set to Geographical North, are unaffected by the deviation within the ship and will tolerate the random motions of a ship at sea.

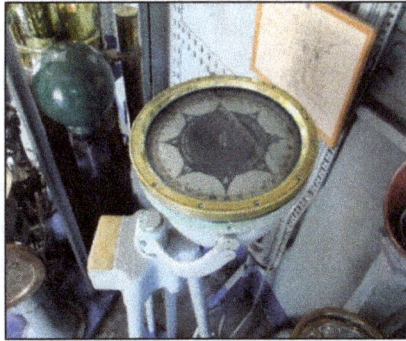

Figures 1.17 & 1.18: The gyrocompass of a WW2 naval frigate (left) and a gyro repeater in the wheelhouse

The orientation of the main gyrocompass relative to its North (i.e. the bearing of the ship) can be transferred by electrical cable to gyro repeaters in the Bridge or any other part of the ship.

Figure 1.19: Getting a bearing with ship's pelorus (Photo: US Navy)

Despite these innovations, Mariners still make use of the magnetic compass in smaller boats, and when inshore will often take bearings off prominent landmarks to check their position on their chart. To do this they have a form of sighting compass mounted on gimbals and having sighting vanes on a rotatable arm on its top surface. With this **pelorus**, the navigator can sight a landmark, take a compass bearing then add or subtract 180^0 to get a back-bearing form the landmark identified on the chart to the ships position. At least three widely spaced and distant landmarks are needed to triangulate back bearings to get an accurate position.

A more modern use of this **triangulation** procedure uses 24 artificial satellites positioned about 20,000 km. above the Earth's surface. They form the **Global Positioning System (GPS)** in which the satellites transmit microwave radio signals which must be picked up by a GPS receiver. The

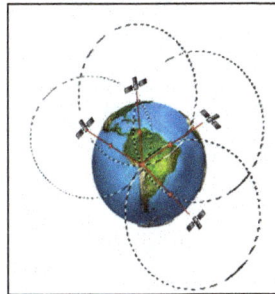

Figure 1.20: A representation of the GPS satellites using triangulation to locate a position on Earth

GPS receivers take this information and use any four of the satellites in range to triangulate and to calculate the user's exact location. The GPS receiver compares the time a signal was transmitted by a satellite with the time it was received. The time difference gives the GPS receiver the distance to the satellite. Now, with distance measurements from several satellites, the receiver can determine the user's position and display it on the unit's electronic map. Even though this system is extremely accurate and widely used in navigation and guidance systems, errors may also occur. These may be due to inaccuracies in the atomic clock of the receivers, the type of map data used, poor reception of satellite data due to the terrain and the range of the satellites, atmospheric problems due to the **ionosphere**, an atmospheric layer which is highly electrically charged and is susceptible to Sunspot interference, and any error which may be added by the military for security purposes.

CHAPTER 2: BECOMING A SCIENCE

2.1 Modern Oceanography

From the middle of the 19th Century, there had been systematic attempts to study the oceans. **Sir James Clark Ross** (English: 1800-1862) had taken bottom soundings of some of the deeps, explored the seas near both poles and had discovered the Earth's Magnetic North Pole. The *Beagle* expeditions with **Charles Darwin**

Figure 2.1: Dredging aboard HMS *Challenger* (Photo: NOAA)

(English: 1809-82) as naturalist and **Robert Fitzroy** (English: 1805-65) as Captain, an able meteorologist and hydrographer in his own right, returned considerable information about the sea and its wildlife. **Matthew Fontaine Maury** (American: 1806-1873) in the United States had also detailed a lifetime of Oceanographic studies in his book *The Physical Geography of the Sea* (1855).

Perhaps the first real scientific approach to the study of the oceans came with the *Challenger* expedition between 1872 and 1876. This extensive expedition was centred on *HMS Challenger*, which had been fitted out by the British Royal Society specifically for the study of the world's oceans. They used the best instruments available at the time, including deep lead lines for sounding the ocean floor, dredges and drag nets to obtain biological specimens and thermometers to measure the water temperatures at different depths. One of their most startling discoveries was what is now known as the Challenger Deep – the deep trench nearly 11,000 metres deep in the southwest Pacific Ocean between Guam and Palau.

2.2 New Techniques

Oceanographic data may be obtained On Station at a specific anchored position or at a shore locality such as a research station, or Underway on board a moving vessel, gathering data along designated courses. Data gathering is made using a wide range of equipment from simple nets, dredges and recording instruments to sophisticated vessels such as submersibles, both manned and unmanned, as well as satellites and other remote sensing devices. Some of the most common instruments used include:

- Plankton nets are used at various stations at sea to catch the very small animals (zooplankton) and plants (phytoplankton) which drift in the upper layers. Sampling can be made at different depths to determine the daily rhythms of vertical migration of these small organisms which provide food for larger animals as well as sediments of external cases and shells upon their death.

Figure 2.2: Towing a plankton net (Photo: NOAA)

- Dredges and grabs are the simplest method of sampling the sea-floor sediments, and have been used in sea chart mapping for centuries.

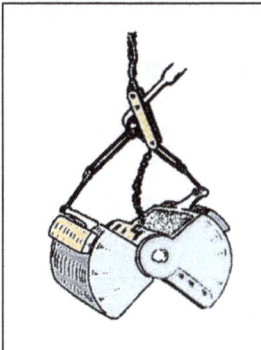

Figure 2.3: A Petersen Grab for sampling bottom sediment

- Corers which can sample the sea floor sediment from a few centimetres to several metres below the floor surface. These consist of a long, hollow weighted pipe with cutting jaws at one end which is forced into the sediment by the weight of its fall. This pushes the sediment up into the pipe and upon return to the ship by reeling the attached cable, the plastic core liner is removed and the sediment then cut into sections for analysis. Piston Corers are a modification which has an internal plunger which operates when a trigger weight preceding the corer hits the sea floor. This causes an internal piston to travel down the body of the coring tube which compacts the sediments inside.

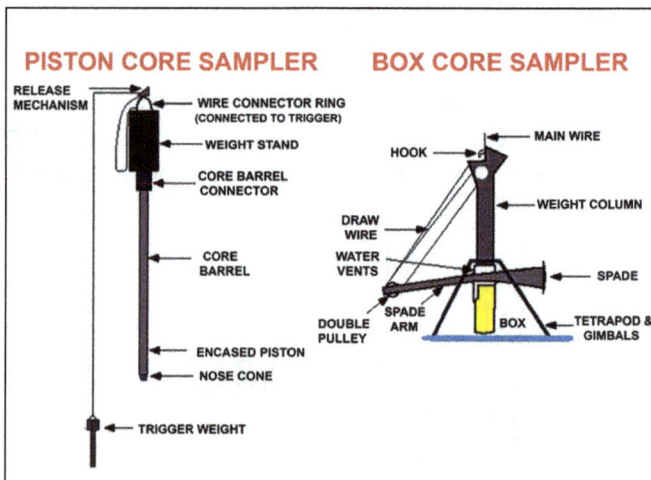

Figure 2.4: Piston and Box Corers (photo: USGS)

- Thermometers and heat-flow probes measure the amount of geothermal energy which flows through the **crust** of the Earth and into the sea. Probes contain small thermistors, transistors sensitive to temperature changes, which are attached to corers and lowered down into deep trenches. Highest heat flows occur over mid-ocean ridges and probable volcanic sources.

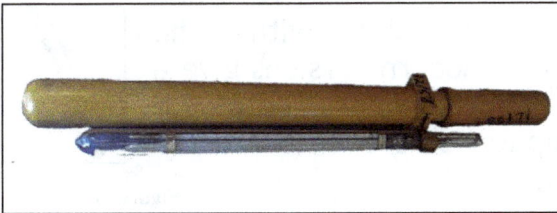

Figure 2.5: An early deep sea thermometer and its wooden case

- **Nansen Bottles** and their successor, the Niskin Bottle, are means by which water samples are obtained at various depths. These containers are closed shut when they reach the required depth then reeled in and the water analyzed for dissolved salts. Special thermometers attached

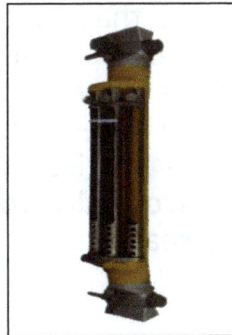

Figure 2.6: An early Nansen Bottle

to the bottles give the temperature and pressure at the sampled depth.

- Mechanical **bathythermographs** are bomb-shaped tubes containing an alcohol thermometer and a barograph which is a water pressure measuring device. Data is automatically recorded on a glass plate within the device. Modern versions give a direct reading onto a computer screen on board the ship.

Figure 2.7: A early mechanical bathythermograph

- **Side-scan sonar** uses fan-shaped, pulsed sound frequencies emitted from a towed sonar device that emits conical or fan-shaped pulses toward the seafloor across a wide angle path across the direction of the ship. The intensity of the reflections from the seafloor is recorded as a series of slices which are then electronically together to form an image of the sea bottom.

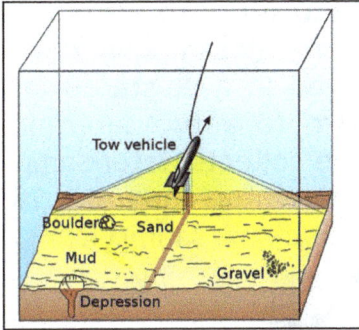

Figures 2.8 & 2.9: Diagram of a side scan sonar device on tow (Left - USGS photo) and the pattern received on-board the ship (Photo: NOAA).

- Marine **magnetometers** are devices which measure the Earth's magnetic field. They may be towed behind a ship at sea or behind an aircraft over land.

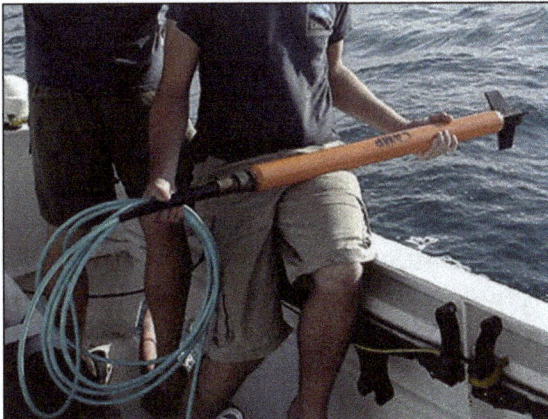

Figure 2.10: Launching a marine magnetometer (Photo: NOAA)

- **Conductivity Temperature Depth Array (CTD)** consists of several vertical sampling devices (Niskin Bottles) arranged in a clustered array and dropped overboard to a pre-determined depth where the device collects water samples and measures sea temperature and electrical conductivity (a measure of the salinity).

Figure 2.11 A CTD array being lowered (Photo: NOAA)

- Deep sea drilling using modified oil-drilling equipment to provide rock data from deep within the sea-floor. Drilling rigs may be anchored as off-shore platforms, as in marine oil rigs, or may be incorporated into special oceanographic ships, such as *Glomar Challenger* (active 1968-83) and its replacement the Joint Oceanographic Institutions for Deep Earth Sampling ship *Resolution*. These ships travel to their designated locations where they hold their position using stabilizing propellers, and send down drill bits on the end of long drill pipes much in the same way as oil rigs on land – except that they often have considerable depth of water above the first point of sediment contact. Once the drilling has reached an appropriate depth into the seafloor, the drill cores are brought up, cut into sections and analyzed on board.

Figure 2.12: JOIDES *Resolution* (Photo: William Crawford, IODP-USIO)

- Remotely-Operated Undersea Vehicles (ROVs) are complex, robotic devices which are launched from a ship under their own power but controlled by a umbilical cable which provides power and receives data from the vehicle. They can perform a wide range of sampling tasks including water clarity, water density, temperature, sound velocity, light penetration as well as obtaining water samples and sea-floor sediment using on-board corers.

Figure 2.13: *Hercules* – one of several types of ROV (Photo: NOAA)

- Manned submersibles are miniature research submarines which can carry crew members for visible observation in deep parts of the ocean as well as for directed sampling.

Figures 2.14 & 2.15: At left, the bathyscaphe *Trieste*, an early deep ocean research submersible which explored the Challenger Deep of the 10,911 metres deep Marianas Trench in 1960 (Photo: US Navy). At right, The highly maneuverable deep sea vehicle *Alvin*, owned by the US Navy but operated by the Woods Hole Oceanographic Institution (Photo: NOAA)

CHAPTER 3: MOVING SEA AND WIND

3.1 Winds - Currents of Air

Winds are caused by the differential heating of the Earth's surface which then heats the gases of the atmosphere above it. In simple terms, the equatorial region will heat up more than the polar regions because it received the Sun's heat radiation directly at ninety degrees to its surface. This causes giant **convection cells** of hot, rising air and cold sinking air. At the Equator then, hot air will rise leaving behind a region of lower air pressure behind. At the poles the colder air will sink and accumulate giving higher air pressure there.

As gases move from regions of high pressure to regions of low pressure, the polar winds should generally flow in towards the Equator. They do so until they sink at about 30^0 north and south as part of the **Hadley Cell** system of the **troposphere**. Here they form another high pressure system of calms and variable winds, the subtropical high, also called the **horse latitudes**. These lie between $30°$ and $38°$ north and south of the Equator and are probably named from the dead horse ceremony when sailors paraded a straw effigy of a horse around the deck before throwing it overboard. This signified the time at which they had earned, by their extended time at sea, their advanced pay and were now free of debt. Another version suggests that the Spaniards, carrying their horses to the New World were forced to kill and throw their horses overboard having run out of feed due to the excessive time without winds in these latitudes. Two smaller cells, the **Ferrel Cell** and the **Polar Cell** also assist in the sinking and rising of air masses in each hemisphere. At the Equator, much of the momentum of the winds has been lost and the region is also a place of low pressure with warm air generally rising, so early mariners found it a place of little or no wind and called it the **doldrums**. Occasionally, winds blowing into this warm, wet climate would bring sudden squalls and storms. This zone is more technically called the **Intertropical Convergence Zone (ITCZ)**, and is the area encircling the earth near the Equator where the northeast and southeast trade winds come together.

Instead of a direct northerly or southerly route, the World's winds are deviated from a straight line by the rotation of the Earth. This is called the **Coriolis Effect** (named after the French Mathematician, **Gaspard-Gustave de Coriolis** (French: 1792 – 1843) which states that moving objects on the surface of the Earth such as wind masses, are deflected to the right with respect to the direction of travel in the Northern Hemisphere and to the left in the Southern Hemisphere. The horizontal deflection effect is greater near the poles and smallest at the Equator, so the winds blowing north in the Southern Hemisphere are deflected to its left, that is towards the West. For example, those winds from Latitude 30^0 South would appear to be coming from the southeast. These are the prevailing southeast trade winds because they were used by mariners to sail their ships to popular trading ports.

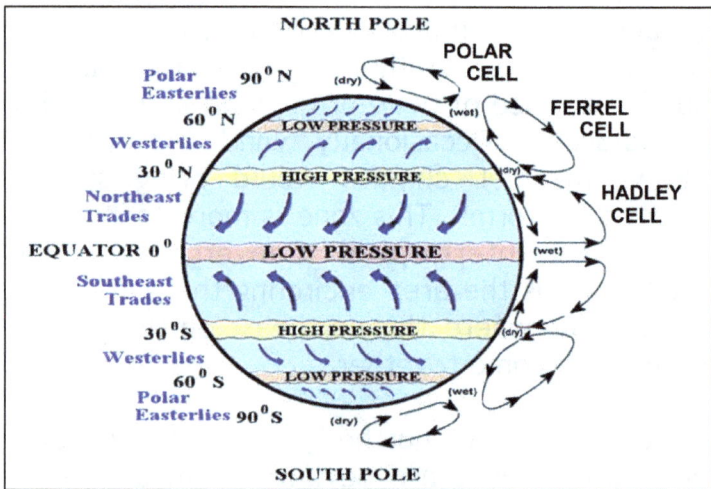

Figure 3.1: Diagram showing the world's wind system

At a local level on a coast, the winds demonstrate these world-wide principles on a smaller scale. During the day, especially in summer, the land will heat up by solar radiation more than the sea. The air above the hotter land will rise creating a slightly lower air pressure than that out to sea. Consequently cool air moves in from the higher pressure above the sea giving a cool sea breeze in the afternoon. During the night, the reverse happens; with the land rapidly losing its heat and the sea retaining much of its heat because of the thermal contents properties of seawater. Consequently there is now warmer, rising air above the sea producing lower air pressure here and slightly cooler, higher pressure over the land. The wind will now blow from the land to the sea as a land breeze.

3.2 Meteorology – The Science of Weather

The winds of the world are thus driven by differences in temperature and pressure and they are also affected by the Earth's rotation and surface features. Winds also carry considerable moisture across the Earth's surface and this becomes vital to rainfall patterns around the globe. It is essential then, for all those concerned with winds and the weather to develop methods of measuring their parameters and develop logical ideas about the patterns observed.

Meteorology is the main science concerned with this measurement and interpretation.

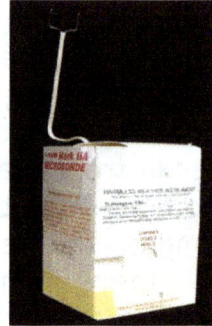

Figures 3.2 & 3.3: A weather balloon being launched (left) and a radiosonde (right) carried by the balloon to measure wind speed & direction, pressure, temperature and relative humidity at different altitudes (Photos: NOAA)

Figure 3.4: A drone is released to monitor environmental conditions (Photo: NOAA)

Figure 3.5: An ocean weather buoy which can measure a wide range of atmospheric and oceanographic parameters (Photos: NOAA)

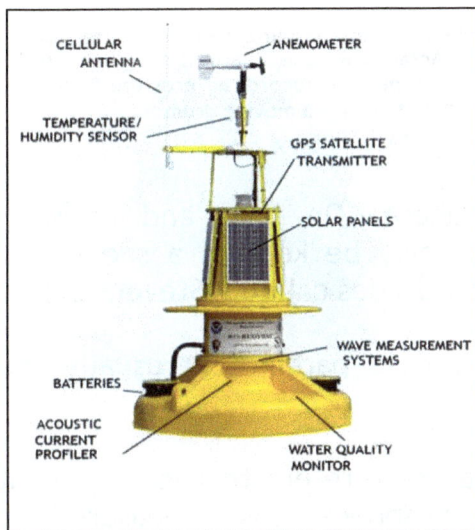

Figure 3.6: Details of a moored weather buoy and the various Instruments used to sample the local environment and then transmit the data via satellite to the receiving stations such as the National Data Buoy Center (NDBC) of NOAA (Photo: NOAA)

There are several instruments used by meteorologists to measure the specific aspects of

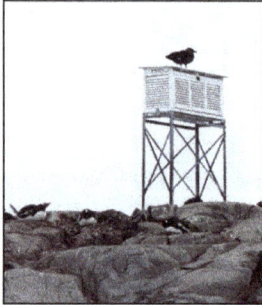

Figure 3.7: Steven Screen, Port Lockroy, Antarctica (note the penguins below and the skua on the box. Skuas love penguin eggs)

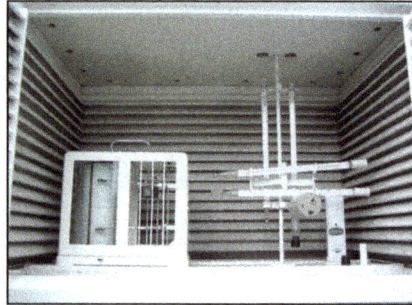

Figure 3.8: Inside the screen – note the Barothermograph (left) which records air pressure & temperature on a moving drum.

the atmosphere. On land and at sea, these instruments may be kept in a special box with open louvered sides called a Stevenson Screen.

The atmospheric parameters usually measured include:

- Temperature – refers to the heat content of the atmosphere as measured by a thermometer in Celsius (C), Fahrenheit (F), or Kelvin (K) degrees. At manned meteorological stations a traditional liquid-filled thermometer is used. This operates on the principle that the liquid expands more that the glass when heated. Red coloured alcohol-filled thermometers are often used in preference to mercury for atmospheric measurements because they can be used down to –115.9 °C (–174.82 °F) whereas mercury freezes at -38°C (-36.4°F) . Their upper limit of +78 °C (172°F) is satisfactory for the intended use of measuring air or sea temperature. Maximum-minimum thermometers are usually U-shaped and filled with mercury and alcohol and have a gas bubble at one end to allow for expansion (they are also called Six's thermometer - named after their English inventor, **James Six: 1731-93**).

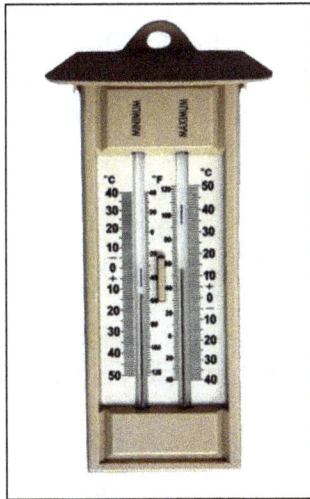

Figure 3.9: A Six's thermometer - as the temperature rises, the mercury expands and pushes a visible, steel indicator up the tube. As the temperature drops, the indicator on the other side of the U-tube stays in its "minimum" position as the mercury drops. Both steel indicators can then be restored to the current temperature using a magnet.

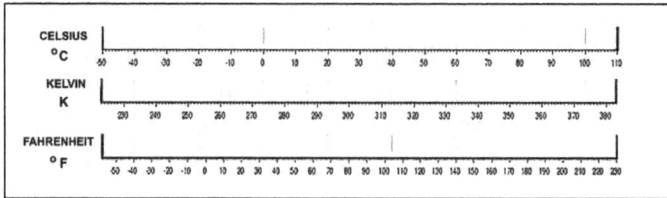

Figure 3.10: (above) Comparisons of the major temperature scales. Conversion from Celsius to Kelvin is given by [K] = [°C] + 273.15 and from Fahrenheit to Celsius by [°C] = ([°F] − 32) × 5/9

Digital electronic devices can also be used to measure temperature, especially in remote locations such as buoys, radiosondes attached to balloons, and unmanned weather stations. Data from these can be stored electronically or transmitted to a central station via satellite. These devices can include **thermistors** which are a type of electrical resistor or a metal alloy wire, dependent on temperature, such as in the resistance temperature detector (RTD) temperature probes used in automatic weather stations.

• Atmospheric pressure is the weight of the atmosphere overhead, that is, the force exerted on a unit area and is measured in newtons per square metre, or pounds per square inch (psi). As well as these common force units, air pressure may also be expressed in millimetres or inches of mercury of the height of a column of mercury supported by the pressure of the air above. This principle was used in 1643 by **Evangelista Torricelli** (Italian: 1608-1647) who described the first mercury barometer. Around 1800, the Fortin's Barometer, named after the instrument maker **Jean Nicolas Fortin** (French: 1750 – 1831), became a practical instrument which could be used at sea and elsewhere for measuring air pressure. Like Torricelli's invention, this consisted of an inverted glass tube, sealed at one end and filled with mercury with the lower open end placed under a pool of mercury in an open bowl. The air pressure above kept the liquid from falling out. Fortin's barometer was more finely made, was housed in a protective metal tube mounted on gimbals and had an adjustable scale which would allow for temperature changes and expansion. Normal air pressure at sea level supports a column of mercury of 760 millimetres or 29.92 inches of mercury. This is equivalent to 15.693 pounds per square inch (psi).

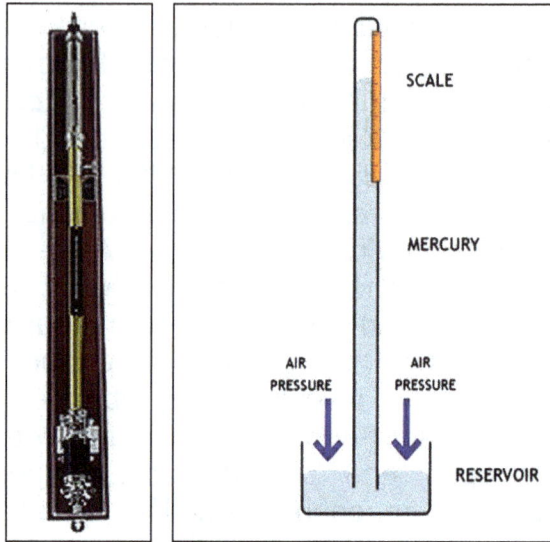

Figures 3.11 & 3.12: Fortin's barometer showing the tall tube of mercury over 900 mm tall (at left) and a schematic view showing its operation (above)

Other standard units preferred in science include: 1013.25 hectopascals (hPa); 1022 millibars; or one Atmosphere. The **pascal** is the S.I. or metric system standard of pressure, and for convenience, the hectopascal (100 pascals) is often taken as being equal to the **millibars (mb)**. Air pressure can also be measured using an **aneroid** barometer, from the Greek *a-* for without and *nēron* for water, as early barometers often used water instead of Mercury but the height of water supported by air pressure would be over 10 metres! The aneroid barometer was invented in 1844 by scientist **Lucien Vidi** (French: 1805 - 1866) as a compact barometer which uses a small, flexible metal box made from an alloy of beryllium and copper. The evacuated box or capsule, or usually several capsules stacked together to combine their variations, is prevented from collapsing by a strong spring. Small changes in external air pressure cause the cells to expand or contract, moving the pointer of the numbered dial by a series of levers.

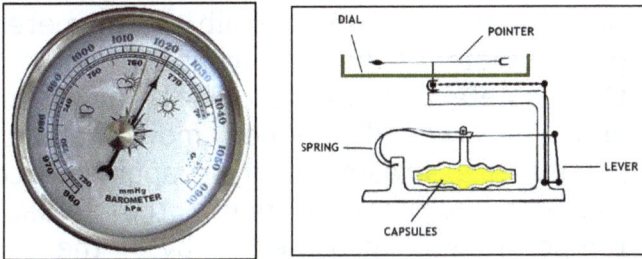

Figures 3.13 & 3.14: An aneroid barometer (at left) and a schematic view showing its operation (at right)

Changes in air pressure signal shifts in the weather, with a sudden drop in pressure suggesting that rain or storms may occur because lower air pressure allows moisture-laden winds to blow into this low pressure area. A rise in air pressure suggests finer weather as the winds would blow outwards taking any moisture with it. Sudden and extreme drops in air pressure at sea level (say past 900 hPa) would indicate an impending storm as the wind would blow inwards at a higher rate and carry more water with it.

- Humidity refers to the amount of water vapour in the air. It is usually expressed as **relative humidity**, or the amount of water vapour which the air contains as a percentage of the maximum amount it can hold at the same temperature. Warm air can hold more water vapour than cooler air. Manually, relative humidity can be measured using hygrometers such as the wet-and-dry-bulb thermometer or psychrometer, which consists of two thermometers, one covered with a wet cloth and the other dry. When moved about or placed in a stream of air, evaporation cools the wet thermometer below the actual air temperature measured by the dry thermometer. Evaporation and cooling depends on how dry the air is at a given temperature. A chart is then used to determine the relative humidity from the amount of cooling. Digital electronic hygrometers can be used for quick readings and in Automatic Weather Stations (AWS) with data logging. Some home versions make use of the effect of moist air on fibres which twist and moves a pointer on a simple dial, but these are not accurate and as best can be used only as an indication of a change in the humidity. The **dew point** is the temperature below which water droplets will form. A large increase in humidity may suggest rain if other conditions are favourable.

A simple wet-and-dry bulb hygrometer can be made by using two simple thermometers attached to board so that their end bulbs are freely open on all sides. Around one of the bulbs is wound some cotton gauze which is fed into a small open bottle of water.

The evaporation from the wet, covered bulb causes the temperature to drop. The amount of evaporation depends upon the humidity. If the humidity is high and the air already has a considerable amount of water vapour in it, there will be little evaporation. If the air contains very little water vapour and the humidity is low, then there will be high evaporation and a bigger drop in temperature on the wet bulb. This all works because heat is needed to turn the liquid water on the wet bulb into gaseous water vapour. This heat is taken out of the surface of the gauze (and the thermometer bulb) and so its temperature drops.

Knowing the dry temperature and the difference between the temperatures of the wet and dry bulbs, the humidity can be determined by consulting a set of tables (see below):

WET-AND-DRY BULB THERMOMETER RELATIVE HUMIDITY (%)

Dry Bulb Temp. °C (°F)	Dry Bulb Temperature minus Wet Bulb Temperature C° (zero difference =100% relative humidity)													
	1 (2°F)	2 (4)	3 (5)	4 (7)	5 (9)	6 (11)	7 (12)	8 (14)	9 (16)	10 (18)	11 (22)	14 (25)	16 (29)	18 (32)
4 (39)	85	70	56											
6 (43)	86	73	60	47	35									
8 (46)	87	75	63	51	39	28	18							
10 (50)	88	77	66	55	44	34	24	15	6					
11 (52)	89	78	67	56	46	36	27	18	9					
12 (54)	89	78	68	58	48	39	29	21	12					
13 (55)	89	79	69	59	50	41	32	22	15	7				
14 (57)	90	79	70	60	51	42	34	26	18	10				
15 (59)	90	80	71	61	53	44	36	27	20	13				
16 (61)	90	81	71	63	54	46	38	30	23	15				
17 (63)	90	81	72	64	55	47	40	32	25	18				
18 (64)	91	82	73	65	57	49	41	34	27	20	6			
19 (66)	91	82	74	65	58	50	43	36	29	22	10			
20 (68)	91	83	74	66	59	51	44	37	31	24	11			
21 (70)	91	83	75	67	60	53	46	39	32	26	15			
22 (72)	92	83	76	68	61	54	47	40	34	28	16	5		
23 (73)	92	84	76	69	62	55	48	42	36	30	19	7		
24 (75)	92	84	77	69	62	56	49	43	37	31	20	9		
25 (77)	92	84	77	70	63	57	50	44	39	33	22	13		
26 (79)	92	85	78	71	64	58	51	46	40	34	23	14	4	
27 (81)	92	85	78	71	65	58	52	47	41	36	26	16	7	
28 (82)	93	85	78	72	65	59	53	48	42	37	27	17	8	
29 (84)	93	86	79	72	66	60	54	49	43	38	29	20	11	
30 (86)	93	86	79	73	67	61	55	50	44	39	30	20	12	4
32 (90)	93	86	80	74	68	62	56	51	46	41	32	23	15	8
34 (93)	93	87	81	75	69	63	58	53	48	43	34	26	18	11
36 (97)	93	87	81	75	70	64	59	54	50	45	26	28	21	14
38 (100)	94	88	82	76	71	65	60	56	51	47	38	31	23	17
40 (104)	94	88	82	77	71	66	62	57	52	48	40	32	25	19
42 (108)	94	88	83	77	72	67	63	58	54	49	42	34	28	21
44 (111)	94	89	83	78	73	68	64	59	55	51	43	36	29	23

Table 3.1: Wet-and-Dry-Bulb Thermometer hygrometer conversion table

For example: from the chart (above), the relative humidity with a dry bulb (i.e. normal temp.) of 24^{oC} and wet bulb of 20^{oC} would give a relative humidity of 69% at 24^{oC}

- Precipitation in the form of rain, snow, hail, sleet or mist, can be measured manually or remotely. Rain is falls as drops 0.5 millimetres in diameter or larger, whereas drizzle and mists consists of droplets smaller than 0.5mm. Rain gauges are often simple metal cylinders with a standardized and graduated glass cylinder inside with a funnel on top. The metal case is then fixed in an upright position open to the sky. Gauges must be emptied and the amount recorded manually at regular intervals. A tipping bucket rain gauge is another instrument which is used to remotely measure precipitation automatically. It contains a dual-chamber mini-collection bucket pivoted like a child's seesaw and located beneath a collection funnel. When rain fills into one side of the bucket it tilts, empties and sends an electronic signal to a data logger or transmitter. Some tipping buckets are equipped with internal heaters to melt snow and frozen rain.

Figure 3.15: A tipping bucket rain gauge within its outer container (Photo: Wiki Commons)

- Wind speed and direction are most desirable parameters in weather measurements and these can be measured in several ways. At manned and in remote stations, they can be measured by an anemometer. This consists of a pivoted weather vane on which is mounted four half-spherical cups on a crossed frame or a single propeller, which moves when the wind blows from any direction. Both the rotation of the cups and the direction of the vane can be measured electronically. Wind speed and direction can also be measured using weather balloons, either tracked by RADAR or from an on-board GPS system.

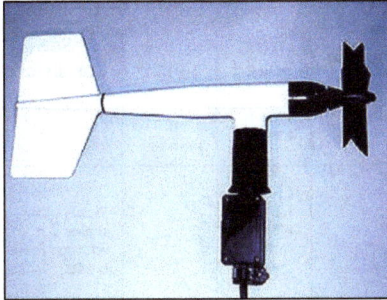

Figures 3.16: A propeller anemometer
(Photo: NOAA)

Figure 3.17: A cup anemometer
(Photo: NOAA)

A most useful system devised for easy observation of wind speed is the Beaufort Scale developed for the Royal Navy by **Sir Francis Beaufort** (Irish: 1774 – 1857) in 1805. This given in Table 3.2 on the next page:

THE BEAUFORT SCALE

Beaufort Number	Description	Wind Speed	Wave Height	Sea Conditions	Land Conditions	Associated Warning Flag
0	Calm	1 km/h 1 knot	0 metres	Sea like a mirror	Calm. Smoke rises vertically.	
1	Light Air	1.5 km/h 1-3 knots	0-0.2 m	Ripples like scales but without foam crests	Smoke drift indicates wind direction. Leaves and wind vanes are stationary.	
2	Light Breeze	6-11 km/h 4-6 knots	0.2-0.5 m	Small wavelets, crests do not break	Wind felt on skin. Leaves rustle. Wind vanes begin to move.	
3	Gentle breeze	12-19 km/h 7-10 knots	0.5-1 m	Large wavelets. Crests begin to break; scattered whitecaps	Leaves and small twigs moving, light flags extended.	
4	Moderate Breeze	20-28 km/h 11-16 knots	1-2 m	Small waves with crests. Frequent whitecaps.	Dust and loose paper raised. Small branches move.	
5	Fresh Breeze	29-38 km/h 17-21 knots	2-3 m	Moderate waves with whitecaps. Some spray.	Branches of a moderate size move. Small trees begin to sway.	
6	Strong Breeze	38-49 km/h 22-27 knots	3-4 m	Long waves with white foam crests frequent. Some airborne spray.	Large branches in motion. Whistling in overhead wires. Umbrella use difficult.	
7	Moderate Gale	50-61 km/h 28-33 knots	4-5.5m	Sea heaps up. Foam blown into streaks Moderate spray	Whole trees in motion. Effort needed to walk against the wind.	
8	Fresh Gale	62-74 km/h 34-40 knots	5.5 to 7.5 m	Moderately high waves breaking crests. Well-marked streaks of foam. Much spray	Twigs broken from trees. Cars veer on road. Progress on foot is hard.	

9	Strong Gale	75-88 km/h 41-47 knots	7-10 m	High waves crests roll over. Dense foam. Large spray reduces visibility	Branches break off trees, small trees blow over. Signs and fences blow over	
10	Storm	89-102 km/h 48-55 knots	9-13.5 m	Very high waves overhanging crests. Large foam gives the sea a white colour. Much tumbling of waves with impact. Large spray reduce visibility	Trees are broken off or uprooted, structural damage likely	
11	Violent Storm	103-117 km/h 56-63 knots	≥ 11.5m	High waves. Very large patches of foam cover sea surface. Large amounts of spray reduce visibility	Widespread vegetation and structural damage likely	
12	Hurricane Force	≥ 118 km/h ≥ 64 knots	≥ 14 m	Huge waves. Sea is completely white with foam and spray. Air is filled with driving spray, greatly reducing visibility	Severe widespread damage to vegetation and structures. Debris and unsecured objects are hurled about	

Table 3.2: The Beaufort Scale

3.3 Layers of the Atmosphere

Studies of the atmosphere using balloon-borne instruments, and later using rockets, showed that the air above was formed in layers.

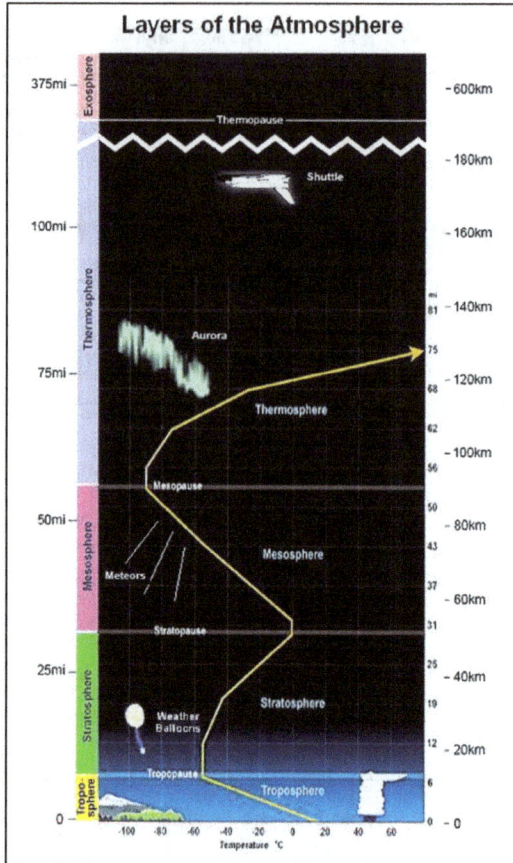

Figure 3.18: Diagram showing the layers of the atmosphere. Most of the wind systems operate in the lower layer, the troposphere. (composite diagram constructed from NOAA images).

The main layers are the:

- **Troposphere** from the Greek: tropos for to mix, and sphaira for sphere, is the lowest portion of Earth's atmosphere. It contains about 75% of the atmosphere's mass and 99% of its water vapour, and it varies in thickness from about 20 km in the tropics to about 7 km at the poles. The troposphere can be divided into two parts: a **Planetary Boundary Layer (PBL),** extending upward from the surface to a height ranging from about 100 to 3000 metres; and the **free atmosphere** above this height.

Figure 3.19: Diagram showing the approximate position of the Planetary Boundary Layer (after NOAA)

The PBL is directly influenced by the Earth's surface at its base by such factors as:

- Frictional drag of the air across the land surface;

- Surface outcrops which add to the frictional drag and impair air circulation;

- Solar heating which heats the land or sea surface and then the air above it; and

- Evaporation and transpiration (water loss from plants) rates.

These factors cause turbulence in the form of eddy currents of various intensities and heights, sometimes as deep as the entire PBL. For example, solar heating of the ground on a sunny day creates thermals of warmer air that rise over colder air causing vertical mixing and turbulence. Naturally most of this frictional shear and turbulence occurs near the Earth's surface with its strength decreasing with height, with less shearing and more buoyancy of the air above.

The PBL is important in meteorology because it is at the Earth's surface where the primary exchanges of heat and water occur and the processes of the world's climate and weather systems occur. Many important weather and climatic phenomena, such as the El Nino/La Nina effects and hurricanes, are principally driven by atmosphere-surface interaction. Surface and boundary layer processes must be identified and understood in order to be used in computer modelling to forecast daily weather, the extent of air pollution events, and the impact of human activity on future climate. These processes are sometimes poorly represented in some current weather models and accurate forecasts and predictions require a careful consideration of PBL processes.

Above the PBL, the free atmosphere of the troposphere is relatively quiet and free of turbulence for most of the time. Here, the wind is almost **geostrophic**, that is, parallel to the **isobars**, the imaginary lines on the weather map joining places of equal air pressure. Within the PBL however, the wind is affected by surface drag and sometimes goes across the direction of the isobars.

In the troposphere, air temperature generally decreases with height, at a lapse rate of about 6.5°C/km until the **tropopause**, or the boundary between the troposphere and overlying stratosphere is reached. This **Environmental Lapse Rate (ELR)** is the difference in temperature between the surface and the tropopause divided by the height. **Jet streams**, fast-flowing, meandering streams of air, usually occur in the upper troposphere just below the tropopause. The tropopause can also act as an inversion layer, where the air temperature ceases to decrease with height and often remains constant. This can trap pollution, such as smog as well as trapping moisture, keeping both close to the ground. If this cap is broken, any moisture present may then build up into violent thunderstorms with freezing rain in cold climates.

- **Stratosphere** from the Greek: stratos referring to its layers or strata, is the second major layer of Earth's atmosphere immediately above the troposphere. It contains about 20% of the atmosphere's mass and has layers of different temperatures, with warmer layers higher and cooler layers closer to the Earth. This increase of temperature with altitude is a result of the absorption of the Sun's ultraviolet radiation by the **ozone (O_3) layer** – a band of air containing a high amount of the unstable form of oxygen molecule ozone which is a molecule having three oxygen atoms. Near the Equator, the stratosphere starts at about 18 km above the surface, whereas at the poles, it starts at about 8 km.

  Temperatures vary within the stratosphere with the seasons, with the greatest variation of temperature, taking place over the poles in the lower stratosphere. Such a temperature profile creates very stable atmospheric conditions, such that the stratosphere does not have the air turbulence prevalent in the troposphere but has strong, steady and mostly horizontal winds. Consequently, the stratosphere is almost completely free of clouds or other forms of weather. The stratosphere is separated from the mesosphere above by the **stratopause.**

- **Mesosphere** from the Greek: meso for middle, occupies the region from about 50 km to 80 km above the surface of the Earth. Temperatures in the mesosphere drop with increasing altitude to about -100^{0}C, making it the coldest of the atmospheric layers. Here, it is cold enough to freeze water vapour into ice clouds called **noctilucent clouds (NLC)** which are most readily visible when the Sun is just below the horizon. The mesosphere is also the layer in which most meteors burn up while entering the Earth's atmosphere. The mesosphere is separated from the thermosphere above by the **mesopause.**

- **Thermosphere** from the Greek: thermos meaning heat is the layer of the atmosphere in which ultraviolet radiation causes **photoionization** of molecules, their dissociation or breakup by light, creating the ions or electrically-charged particles, in the ionosphere. This is a region of Earth's upper atmosphere, from about 60 km to 1,000 km altitude which includes the thermosphere, parts of the mesosphere and exosphere. This ionization is caused by solar radiation and forms the inner edge of the **magnetosphere,** the magnetic field around the Earth. The ionosphere influences long-distance radio transmission around the Earth.

The thermosphere begins about 85 kilometres above the Earth and here the atmospheric gases form into layers according to their molecular mass. Temperatures increase with altitude due to absorption solar radiation and can rise to 2,500 °C during the day. In the **anacoustic zone**, over 160 kilometres above the surface, the air pressure is so low that sound waves do not travel This name is derived from the Greek -ana meaning not and akoustikos meaning to hear. Here too, are atmospheric tides, caused by changes in daily heating producing waves of air which slowly dissipate due to molecular collisions with hot, ionized gases, or plasma in the zone. Many low-orbit space craft, such as the Space Shuttle, travelled within the upper thermosphere just below its upper boundary, the **thermopause** and the

- **Exosphere** from Greek éxō for outside, is the thin, low-density layer of the upper atmosphere. Here, molecules and other particles are still bound to the Earth by gravity, but where the density is too low for them to behave as a gas by colliding with each other. The most common molecules within Earth's exosphere are those of the lightest gases, mainly hydrogen with some helium, carbon dioxide, and a little atomic oxygen near its base. It is often difficult to define the boundary between the exosphere and true outer space.

3.4 More about Surface Winds

The differences in the heating of the Earth's surface, its landforms and the dynamics of the layers of the atmosphere, produce a complex but predictable pattern of wind systems of great convection cells covering the surface and extending into the upper atmosphere.

In tropical regions, from about 30^0 S and 30^0 N, the summers are often the time for rapid development of severe storms. These are generally known as **tropical cyclones** because of their high winds which rotate about an eye of calm. In Southeast Asia they are called typhoons, from the Greek Typhon, named after a monster from mythology but also similar to the Chinese *táifēng* for wind. In the Caribbean, they are known as hurricanes from the Spanish word *huracán* named for the Carib Indian storm god, *Juracán*. Usually cyclones occur in the hottest months of late summer, but other regions experience distinct cyclone seasons. For example, in northern Australia and the southern Pacific they occur from November to April, and in the Northern Atlantic they occur from June to

November. At these times, there is a greater difference in temperature between the sea and the air above. The warmer water experiences rapid evaporation with the water vapour rising in a circular motion as described by the **Coriolis Effect** which is due to the Earth's rotation. These currents rotate clockwise as seen from above in the Southern Hemisphere and anti-clockwise in the Northern Hemisphere. The rapid rise causes an intense low pressure at the surface and a massive build-up of storm clouds with high winds often over 100 kilometres per hour. Torrential rain is common and **storm surges** sometimes occur when the high winds push the surface of the sea outwards. When these surges hit coastlines they often cause localized flooding. These events occur around the central part of the tropical cyclone which experiences relative calm.

Online Video 3.1: <u>Sail</u> a windjammer down the eastern Australian coast from Brisbane to Sydney.
Go to https://www.youtube.com/watch?v=ohaVjUrLN3M

Figures 3.20 & 3.21: At left, TC "Isabel" nearing Florida in 2003 rotating ant-clockwise (NASA photo) but in the Southern Hemisphere cyclones rotate clockwise (at right TC "Aivu" in north eastern Australia, 1989 – NOAA photo).

CATEGORY	STRONGEST GUST (Km/hr)	TYPICAL EFFECTS
1. Tropical Cyclone	Gales - winds less than 125 km/hr	Minimal house damage, some damage to crops, trees and caravans. Boats may drag moorings.
2. Tropical Cyclone	Destructive winds 125-164 km/hr	Minor house damage and significant damage to signs, trees and caravans. Heavy damage to some crops, risk of power failure and small boats break moorings.
3. Severe Tropical Cyclone	Very destructive winds 165-224 km/hr	Some roof and structural damage and caravans destroyed. Power failure likely.
4. Severe Tropical Cyclone	Very destructive winds 225-279 km/hr	Significant roof and structural damage and caravans will be destroyed or blown away. Dangerous flying debris and widespread power failure
5. Severe Tropical Cyclone	Extremely destructive winds more than 280 km/hr	Extremely dangerous with widespread destruction

Table 3.3 Classification of tropical cyclones around Australia

Whether living in cyclone-prone regions or just visiting in the cyclone season, it is prudent to know of some of the safety precautions which would needed when a cyclone strikes:

WHEN A CYCLONE (HURRICANE) STRIKES
• Disconnect all electrical appliances. Listen to your battery radio for updates. • Stay inside and shelter well clear of windows in the strongest part of the building i.e. bathroom, internal hallway or cellar. Keep evacuation and emergency kits (food, water for two days, battery radio + spare batteries, torch, First Aid kit) with you. • If the building starts to break up, protect yourself with mattresses, rugs or blankets under a strong table or bench or hold onto a solid fixture, e.g. a water pipe. • Beware the calm eye of the storm if the wind drops. Don't assume that the cyclone is over, violent winds will soon resume from another direction. Wait for the official 'all clear'. • If driving, stop put the handbreak on and the car in gear. Stay well away from trees, power lines, streams and beaches. Stay in the vehicle unless there is a sturdy shelter nearby.

Figure 3.22: Emergency action in a cyclone

Weather details are often given as **synoptic charts** or weather maps, as RADAR images for rainfall for local regions in the media, on internet weather sites, and on weather smartphone apps. A typical synoptic chart is shown below:

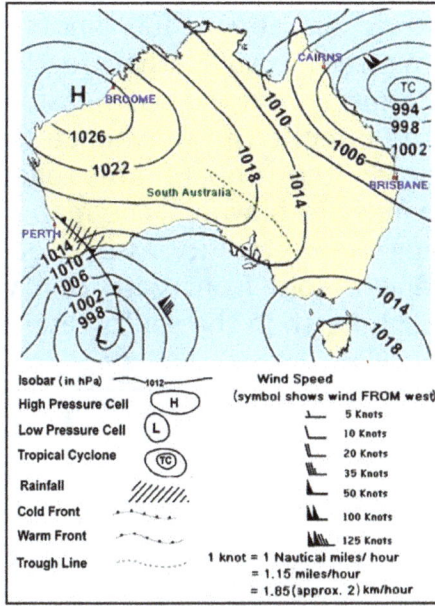

Figure 3.23: Synoptic chart for Australia (Southern Hemisphere system showing a tropical cyclone (TC) in the northeast and other typical chart symbols.

The synoptic chart here is for Australia, and it shows a tropical cyclone (TC) rotating clockwise in the Southern Hemisphere, as a compacted pattern of **isobars.** Lower air pressure, say below 1022 hPa, usually means bad weather near the sea as winds are able to blow inwards from highs to lows, bringing rain. High pressure zones usually give fine weather as winds generally blow outwards. Winds usually follow the isobars directions, so that at the city of Cairns, shown on the map above, there are strong south westerly cyclonic winds at about 100 knots, whereas Broome in Western Australia has gentle winds of only about 5 knots from the southeast. Winds are named by the direction from whence they come, and the closer the isobars are together, the faster the winds are blowing. The map also shows a weak **trough** of instability associated with the two Lows and a cold front with rain which has passed through Perth in the southwest.

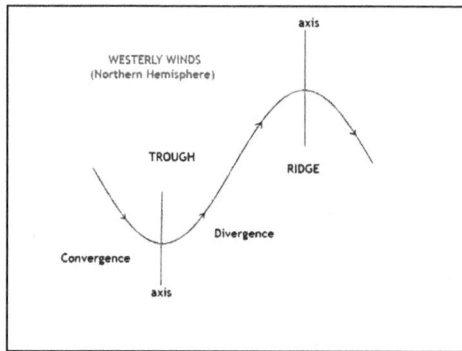

Figure 3.24: Diagram showing a trough-ridge system

To understand the weather pattern, it should be remembered that it operates as a three-dimensional system. Whilst lows and highs generally concern surface conditions, they do extend upwards to a limited extent into the troposphere. Troughs and **ridges** also form as these pressure areas are stretched and extended across the surface by conditions within the planetary boundary layer. A trough is an elongated region of relatively low atmospheric pressure, where the air has cooled then sunk, whereas a ridge is an extended region rising air and high pressure. If a trough forms in the mid-latitudes, a temperature difference between two sides of the trough usually forms as a **weather front.** A weather front here is usually less convective, i.e. less movement of air vertically than in the tropics or subtropics where sudden rising air masses can cause hurricanes and cyclones. Unlike fronts, there is not a universal symbol for a trough on a weather chart, but in some countries they marked as a dotted line (see Figure 3.24), a dashed line or solid line between low pressure regions.

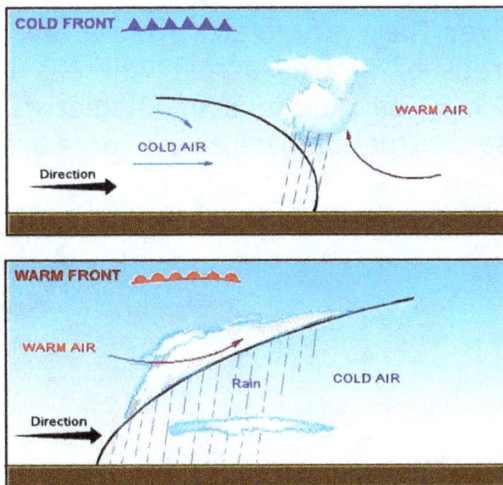

Figure 3.25: Diagrams showing advancing cold fronts (top) and warm fronts (bottom)

In the mid-latitude westerly winds, troughs and ridges often alternate, especially when upper-level winds are in a steep trough-ridge pattern. For a trough in the westerlies, the region just west of the trough axis is typically an area of convergent winds (coming in) and descending air - and hence high pressure - while the region just east of the trough axis is an area of fast, divergent winds (going out) and low pressure.

In parts of North America, Argentina, Western Europe, South Africa and Australia, **tornadoes** also form when sudden and severe storms activity occurs. Tornadoes are mainly land-based, vortexes of rotating air which are very much smaller in size than tropical cyclones but very destructive. They vary in size from a few metres to hundreds of metres across whereas tropical cyclones may be hundreds of kilometres across and cause destruction over a very large area.

Figure 3.26: A tornado in Oklahoma, USA (Photo: NOAA)

Tornadoes are due to the rapid solar heating of the land surface causing an upward spiral rush of air associated with a single storm, and are relatively short-lived. Over water, on lakes and at sea, they form water spouts.

Another major climatic process involving air masses is the **monsoon system** of Asia, especially around the Indian Ocean. A monsoon is a seasonal change in the direction of the prevailing winds and this can cause wet and dry seasons throughout much of the tropics. Monsoons always blow from cold to warm regions because the land heats up more than the sea and this hot air rises causing a low pressure zone which causes the winds to blow in from the sea, bring warm, moist air and thus considerable rains, such as the summer monsoon of India which occurs around April to September. A dry winter monsoon also brings winds with little moisture from the northeast from central China with most of the moisture being removed by the Himalaya Mountains from October to April. In the Southern Hemisphere, the wet monsoon season lasts from September through to February over Indonesia, the Philippines, Irian Jaya and northern Australia with north-easterly winds blowing in from Southeast Asia.

3.5 Clouds

The names comes from the old English word clud, meaning a hill or a mass of rock, and are formations of accumulated moisture droplets or ice crystals, within the atmosphere That is, they are **aerosols** of droplets within a gas. They are generally formed as moist air rises into the cooler atmosphere, mainly within the troposphere but some high altitude clouds can also form in the stratosphere and mesosphere.

The modern terms using in naming clouds developed from the system introduced by **Luke Howard** (English: 1772 – 1864) in 1802. In the modern system, clouds are defined by descriptive terms in five categories:

- Stratiform or sheets

- Stratocumuliform or rolls and ripples

- Cirriform or wisps and patches

- Cumuliform or heaps of variable size and

- Cumulonimbiform or very large heaps with complex structure.

In addition, high altitude clouds may have the prefix -alto added to their name to indicate their superior height. The appearance of clouds has long been used by sailors, farmers and later, meteorologists for foretelling future weather conditions as they indicate the processes which produce weather changes. Much of this also involves local conditions. For example, on the east coast of Australia, a sudden build-up of tall 'thunderheads' of cumulonimbus clouds suggest that storms may be approaching. This is especially true if they are seen forming in the south in front of a 'Southerly Buster' or sudden influx of colder, moist wind from the south. High altitude, wispy clouds of alto cirrus clouds will often herald rain within 48 hours.

Figure 3.27: Thunderheads building up above Port Denarau, Fiji.

Details of the most common cloud types and their associated weather patterns are given below:

HIGH ALTITUDE		
Cirrus	**Cirrostratus** Often form as an	**Cirrocumulus**
These are curved, fibrous clouds which are often called "Mare's Tails". They have a transparent appearance and are composed of ice crystals. In some parts, their appearance often means wind and rain within a few days.	extensive, sheet over much of the sky with the Sun often appearing as a dim halo due to the refraction by ice crystals. They also may indicate the approach of a Warm Front and rain.	Patchy and broken, these clouds may also show ripples of high (5 Km.) winds. They also contain ice crystals in long clumps. They which may indicate fine weather if there has been rain.
MEDIUM ALTITUDE		
Altostratus	**Altocumulus**	**Nimbostratus**
Generally a blue of grey layer covering most of the sky. Semi-transparent, there is no halo around the Sun, and may form ahead of a Warm Front and give light rain.	Often appears as sheets with many ripples or rolls or patches due to re-working by high winds. Often called a "Mackerel Sky" and usually means fine weather	Form as a continual, dark grey rain cloud covering the entire sky and giving light to moderate rain over an extended period.

LOW ALTITUDE		
Cumulus	**Stratus**	**Cumulonimbus**
Detached dense mounds separated by open sky. They often build up during the day as water vapour rises, particularly after the passing of a Cold Front in a High Pressure zone. They may continue to build up to form storm clouds.	A thick covering of grey cloud with a flat base and may give light rain. Caused by a rising mass of warm, moist air, these clouds may resemble a high fog.	Large thunderheads which build up on hot, moist days with rapid uptake of moisture. They may produce storms with rain and hail and sometimes tornadoes
	Stratocumulus  Dark, rounded masses, usually in separated clumps typical of Sub-tropical and Polar regions. Little rain.	

Figure 3.28: Some of the common cloud types

3.6 Ocean Currents

Ocean Currents are caused by a number of factors including:

- prevailing wind direction
- salinity
- temperature differences
- the Coriolis Effect of the rotating earth
- ocean-floor contours and
- restrictions by landmasses.

Ocean currents also vary in their velocity, salinity, amount of nutrients and temperature. For example, the **Gulf Stream** carries warm water from the tropical regions of the Gulf of Mexico, up the southern coast of the United States and then moving northeast across the North Atlantic to Europe. This gives countries such as Britain, Ireland and Iceland a much warmer climate due to the warming of the atmosphere above the current. Conversely, the east coast of Canada is missed by the warm current and has a much colder climate than these other countries far to the north. Another example is Lima, Peru where the climate is cooler or sub-tropical, than the tropical latitudes in which the area is located, due to the effect of the Peru/Chile (or Humboldt) Current. This current, whilst cold, is also rich in nutrients from the south polar region which makes the region a rich fishing zone.

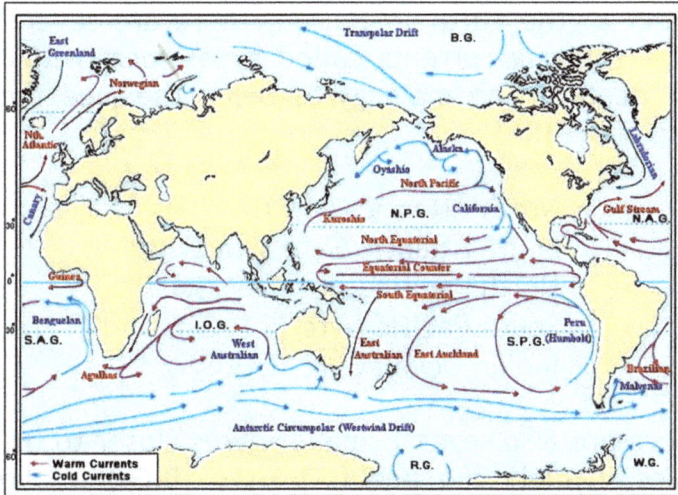

Figure 3.29: Diagram showing the main ocean surface currents.

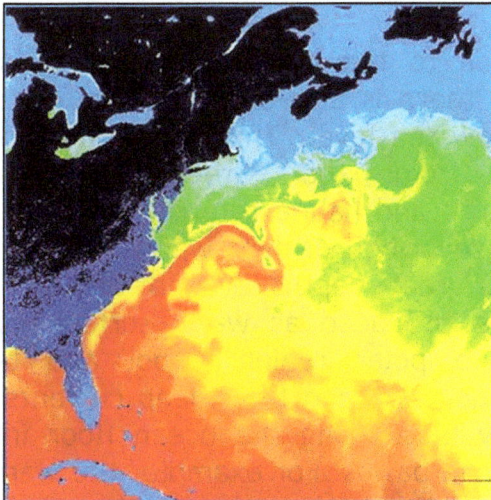

Figure 3.30: Infra-red (heat) image of the Gulf Stream along the east coast of the USA using the Moderate-Resolution Imaging Spectroradiometer (MODIS)onboard the Aqua Satellite (Photo: NASA)

Because of the main currents and the Coriolis Effect of the Earth's rotation, there are several large rotating currents called **gyres** form within the oceans. The main gyres being the (see the map above):

- North Atlantic Gyre (N.A.G);
- South Atlantic Gyre (S.A.G.);
- North Pacific Gyre (N.P.G.);
- South Pacific Gyre (S.P.G.); and
- Indian Ocean Gyre (I.O.G.).

There are also several smaller gyres closer to the Poles, namely the Beaufort Gyre (B.G.) in the Arctic Ocean, and the Ross Gyre (R.G.) and the Weddell Gyre (W.G.) near Antarctica.

In addition to the complex system of surface currents, there is also a sub-surface circulation deep below sea level. This is the **thermohaline system**, from the Greek: thermos for temperature and haline referring to salt. It is a great conveyor belt of deep water currents which travel at great depths around the sea floor. These are produced as warm surface currents travel into polar regions, cool and become denser, then sink and form deep currents which then flow back along the ocean floor into the major ocean basins. For example, the warm Gulf Stream flows northwards across the North Atlantic, giving warmer air temperature to the coasts of western Europe but then cools when it reaches the Arctic Ocean. Here it sinks and flows at depth south under the Gulf Stream and then into the South Atlantic and east between Antarctica and the Africa where it turns north, warms in the tropical region off the east coast of Africa, becoming less dense and rises as a surface current.

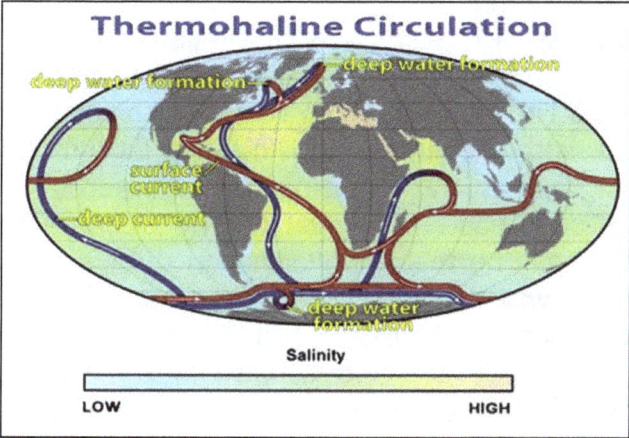

Figure 3.31: Diagram showing the main ocean surface and sub-surface currents (Photo: NASA)

The **El Niño** effect is the climatic event due to the movement of the warm pool of water from the central Pacific across to its eastern side due to a weakening of the trade winds. it is named from the Spanish for the [boy] child because this event often occurs around Christmas when the child Jesus is celebrated in South America. Normally, the trade winds push warm water westward across the Pacific bringing warm, moist air and rain. In the east along the coasts of the Americas, the air is relatively dry and cool due to the up-welling of cold water from the Peru (Humboldt) Current coming from Antarctic waters. When the trade winds decrease, this warm mass of air moves eastward and brings rain and storms to the Americas and reduces the rainfall in the west to countries such as Australia and those of eastern Asia, often causing draught.

The **La Niña** effect, from the Spanish for the girl child, is the reverse of El Niño. In a La Niña event the sea surface temperature across the equatorial eastern central Pacific Ocean will be lower than normal and warmer in the west as the trade winds and the Equatorial Current strengthens. This usually brings wetter conditions in Australia, Asia and usually in the American Midwest, but draught in southwestern USA and South America.

Usually there is a to-and-fro movement of atmospheric pressure within the tropics called the **Southern Oscillation** which produces an alternation of El Niño - La Niña events approximately every three to seven years. The strength of this southern oscillation is measured by the **Southern Oscillation Index (SOI)**. The SOI is computed from fluctuations in the surface air pressure difference between Tahiti in the Pacific, and Darwin, Australia on the Indian Ocean. Sustained <u>negative</u> values of the SOI, say below −8, often indicate an El Niño episode. These negative values are usually accompanied by sustained warming of the central and eastern tropical Pacific Ocean, a decrease in the strength of the Pacific Trade Winds, and a reduction in winter and spring rainfall over much of eastern and northern Australia and Asia.

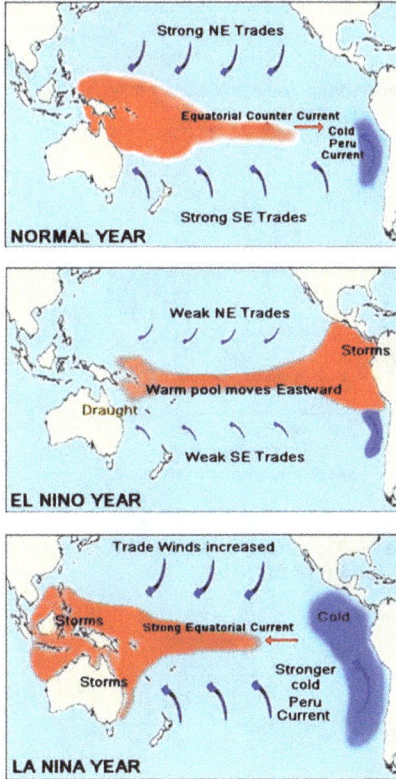

Figure 3.32: El Nino and La Nina oceanic conditions

Sustained <u>positive</u> values of the SOI, say above +8, indicate a La Niña episode, and are associated with stronger Trade Winds across the Pacific and warmer sea temperatures north of Australia. Waters in the central and eastern tropical Pacific Ocean become cooler during this time with the combined effect of giving a wetter climate in eastern and northern Australia and Asia.

Standardized Southern Oscillation Index (SOI)

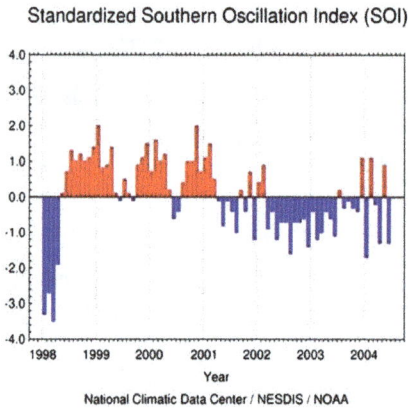

National Climatic Data Center / NESDIS / NOAA

Figure 3.33: Chart showing the SOI for 1998-2004 (Photo: NOAA)

CHAPTER 4: THE OCEAN FLOOR and BELOW

4.1 The Ocean Floor

The discoveries of the 19th and early 20th Centuries had given an obscured picture of a static, flat and barren ocean floor of fine sediments, occasional dramatic underwater mountain ranges, isolated volcanic peaks, some still active, and a few ocean deeps. Very little was known about the depths of the oceans beyond the **continental shelf** – the shallows surrounding most continents. At some locations there are submarine canyons on the outer edge of the continental shelf. Sediments often accumulate here and may then be collapse down slope as a **turbidity current**, a flow of shelf sediment and water accompanied often by fragments of the shelf as well. The low-angle shelf drops off to only a slightly steeper **continental slope** across which finer sediment will flow and be deposited. Further down there is a gradual increase in slope called the **continental rise** which then extends out onto the extensive and flat **abyssal plain** which makes up most of the sea-floor. Here, at about 4500 metres depth, and deeper in the tropics, is the zone where calcium carbonate, used by many sea animals to make their protective coverings, dissolves. This is the **Calcium Carbonate Compensation Depth (CCD).** This depth varies at times, and it is thought that when carbon dioxide levels are higher in the atmosphere, the CCD becomes shallower. This could have implications for corals, molluscs and other shelled animals. In the Eocene Epoch, lasting from 56 to 33.9 million years ago, there was a deepening of the CCD as the climate moved from greenhouse conditions to an ice age atmosphere.

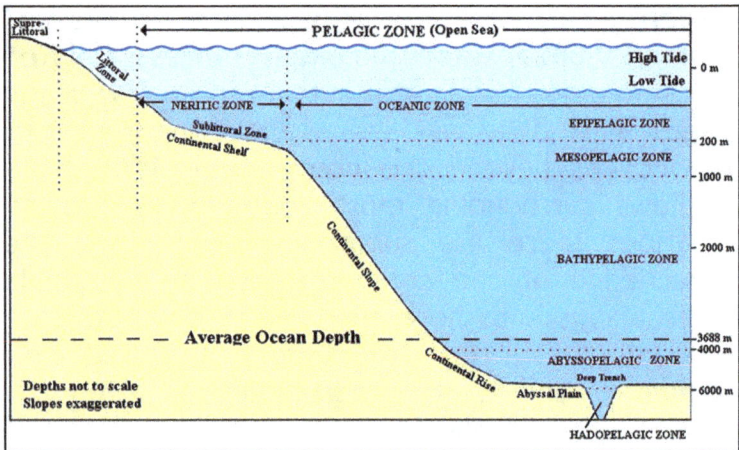

Figure 4.1: Diagram showing the continental margins and depths

By the twentieth century, it was known that the average ocean depth was about 4000 metres and that the deepest part of the ocean was the Challenger Deep in the Marianas Trench system, south of Japan at about 11 kilometres below sea level. Some of the major deeps are shown in Table 4.1:

	TRENCH	DEPTH (m)
1.	Marianas (Challenger Deep)	11,022
2.	Tonga	10,882
3.	Kiril	10,542
4.	Philippine (Cape Johnson Deep)	10,497
5.	Kermadec	10.047
6.	Japan	9,800
7.	New Hebrides	9,165
8.	Puerto Rico	8,385
9.	Bougainville	8,310
10.	Peru-Chile	8,055
11.	Aleutian	7,500
12.	Indonesian	7,450
13.	Aru Trough	3,652
14.	Timor Trough	3,276

Table 4.1: Major Ocean Trench systems

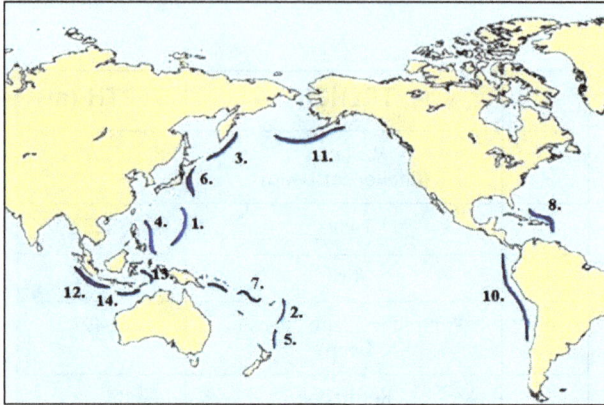

Figure 4.2: Map showing the major deep ocean trenches

Marine sediments on the ocean floor consists of insoluble rock and soil material, transported from the land to the ocean by wind, ice and rivers, as well as the remains of marine organisms, products of submarine volcanism, chemical precipitates from seawater, and material from meteorites which accumulate on the seafloor. Sediments deposited near the continents cover about 25% of the seafloor, but account for roughly 90% by volume of all marine sediments.

Turbidites from turbidity currents off the Continental Shelf are also deposited at some distance across the Abyssal Plains. These consist mainly of a mixture of sands, silts and some gravels and often the remains of organic sedimentary formations such as coral reefs which have been broken off from the Continental Shelf.

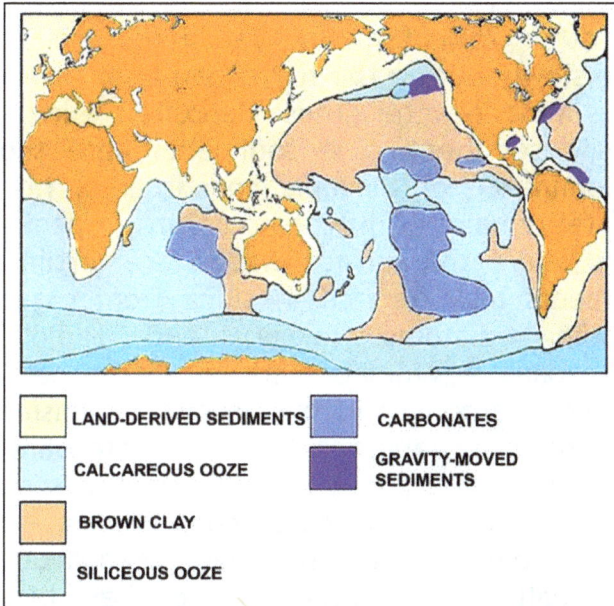

Figure 4.3: A simplified map showing the major types of sea-floor sediments

The sediments deposited on the Continental Shelves and Rises usually settle rapidly and consist of sediments which are indicative of their source (or provenance). For example, carbonate-rich muds often come from nearby coral reefs in tropical regions and silicate glasses and fine volcanic rock fragments come from nearby volcanic sources. Sediments formed in these regions contain both organically-formed sediments and sediments derived from the land and are called **hemipelagic sediments.**

The other 75% of marine sediments accumulate more slowly out on the flat ocean floor of Abyssal Plains in deep water. These deposits are known as **pelagic sediments** and sediments here show a predominance of biogenic constituents, i.e. the skeletal remains of marine organisms, especially in areas where the surface waters are fertile. The sediments may be rich in silica from organisms including diatoms (algae) and radiolarians (Protozoans – animals with a single cell), or calcium carbonate from such organisms as foraminiferans (another type of Protozoan) and pteropods, larger, free-swimming animals similar to sea snails. If the biological constituents are over 30% by volume, then the deep-ocean sediments are usually named according to their biogenic components. For example, a mud containing 30% of foraminiferal tests (shells), may be called a foraminiferal mud or **ooze**. If one specific type of organism dominates, it is may be referred to by its generic name, such as in a *Globigerina* Ooze. Where biogenic constituents are less than 30% of the total composition, the deposit is called deep-sea clay, and reference is made to its colour e.g. a brown mud, or red clay. Usually pelagic sediments remain undisturbed because of the lack of bottom currents but they may be re-worked if the sea-floor suffers from major disturbance from tectonic or volcanic activity.

4.2 Coral Reefs and Atolls

One of the most interesting observations was that by **Charles Darwin** (English: 1809-1882) about the many coral islands and atolls which he had observed during his voyages in *HMS Beagle* (1831-36). He noted that in tropical waters, coral reefs often formed around volcanic islands which varied greatly from large, complex landmasses to places where the central landmass had disappeared completely leaving only the surrounding reefs as an **atoll**. There are three types of coral reef:

- **Fringing reefs** which are attached to a shoreline, usually an island, small or large, with little or no lagoon and a poorly developed Reef Crests. Coral Knolls are not fully developed and the reef will be relatively flat.

Figure 4.4: Fringing reef at Lifou, a coral and limestone island in the Loyalty Group, western Pacific

Online Video 3.2: Dive on the fringing reef at Lifou, Loyalty Islands.
Go to https://www.youtube.com/watch?v=sAf2Ry5DgbE

- **Barrier reefs** which are usually more extensive, being attached to a long shoreline and may consist of a complex of many smaller reefs and associated channels and islands which offer a barrier to the mainland. The Great Barrier Reef off the Queensland coast, Australia, is such an example and it runs parallel to the coast for over 2300 kilometres. Smaller Barrier Reefs exist off other large islands. They usually have a well-developed and deep lagoon or inland waterway with a pronounced outer reef which drops off to the deep ocean and

Online Video 3.3: Sail a schooner to the outer reef of the Great Barrier Reef, Australia
Go to https://www.youtube.com/watch?v=c9NG-R9046Q

- Atolls which are coral reefs, often with groups of very small islands forming a circular pattern around a central lagoon.

Figure 4.5: A satellite photograph of an atoll, Cocos (Keeling) Islands in the Indian Ocean (NASA Photo)

Coral Reefs can have a complex structure as they built up at different locations by the coral polys and other marine organisms which live in the reef structure.

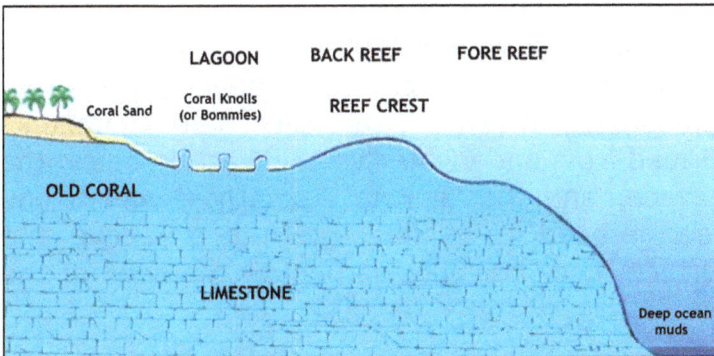

LAGOON   BACK REEF   FORE REEF

Coral Sand   Coral Knolls (or Bommies)   REEF CREST

OLD CORAL

LIMESTONE

Deep ocean muds

Figure 4.6: Diagram showing a simplified structure of a fringing coral reef.

Charles Darwin proposed that atolls where formed from fringing reefs which originally surrounded a central volcanic island which then slowly sank. This allowed the coral to build vertically as the island sunk, so that eventually there was no trace of the central island, only the lagoon which replaced it.

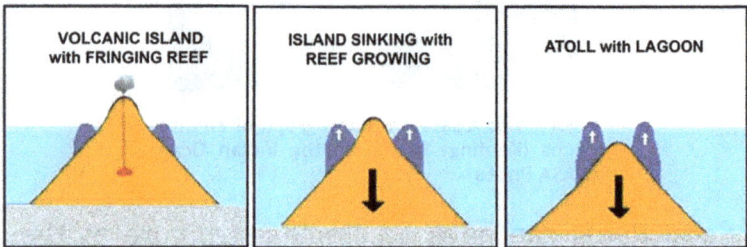

Figure 4.7: Diagram showing Darwin's Theory of Atoll Formation – as the island sinks and the reef grows, the central land component is replaced by a lagoon.

There was also a similar solution for the formation of **guyots,** which are flat-topped submarine seamounts. It was suggested that seamounts formed from underwater volcanic action below the surface, and that guyots were those which had been eroded to a flat level at or above the surface, and then subsided below the level of the sea.

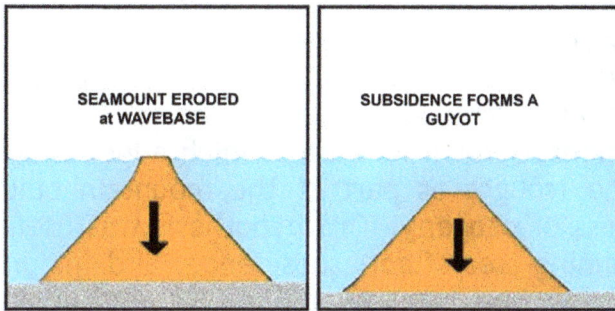

Figure 4.8: Diagram showing the formation of a guyot

4.3 New Discoveries

The concept that parts of the sea floor could slowly subside was not a new idea, but as yet there was not a good explanation for it. In the mid-19th Century, the American geologists **James Hall** (1811-98) and **James Dwight Dana** (1813-95), suggested the **geosyncline hypothesis.** This suggested that parts of the oceans had subsided as long troughs, which they called **geosynclines**, due to the weight of sediment being washed into them from the land. As the heavy trough sank into the Earth's crust, it became folded from each side due to the lateral forces due shrinking effects of the Earth as a whole. As they sank, the bases of these geosynclines would fold, heat up and melt forming pockets of molten rock, or magma, which would intrude upwards due to its lighter density. The overall effect would be that the whole geosyncline would be folded up above sea level becoming a new mountain chain so as to maintain the balance of **isostasy**. This was the balance in equilibrium between erosion of mountains which remove material from the upper crust, and the sedimentation within basins which added material to the troughs as part of the mountain building process of **orogeny**. This hypothesis was useful in explaining sea floor subsidence and mountain building, but by the 1960's it was generally considered obsolete.

After the end of the Second World War, new advances in underwater techniques, following on from submarine warfare, the development of sonar (an acronym for **SO**und **N**avigation **A**nd **R**anging), and the desire to learn more about what was below the sea which covered more than 70% of the Earth's surface. This led to new research with the revision of some earlier ideas, and some startling new observations concerning:

- Continental fit or the concept that the continents had once been fitted together but had later moved apart. This was not a new idea, having been suggested by **Abraham Ortelius** (Flemish: 1527-1598) in 1596, **Alexander von Humbolt** (Prussian: 1769-1859) in 1801, **James Dwight Dana** (American: 1842-1895) in1849, and **Eduard Suess** (Austrian: 1831-1914) in 1861. Suess had proposed the breakup of a super continent, which he called Gondwana or Gondwanaland after the Gondwana region of central northern India which is derived from

Sanskrit for "forest of the <u>Gonds</u>" – the people who live there. **Alfred Wegener** (German: 1880-1930), a geophysicist and meteorologist, was also unsatisfied with the geosyncline hypothesis, and saw merit in this suggestion that the continents were able to move apart. He knew of the widely recognized fact that Africa and South America appeared to fit together like jigsaw puzzle pieces, and so he collected information about the climate of these areas in the geologic past as recorded in rocks. He recognized belts of coal crossing from North America in Europe and Asia, and found evidence that an ice sheet had once advanced across southern Africa and India, a phenomenon that was impossible to explain in the modern arrangement of the continents. His conclusion was that the modern continents had been formed from one great super continent which had broken apart and moved to their present location.

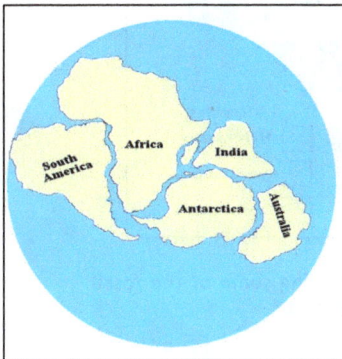

Figure 4.9: Diagram showing a probable fit of the Southern Hemisphere continents. Note that India (now in the Northern hemisphere) is included here.

- Fossil evidence also suggested that the continents once fitted together. Plants such as *Glossopteris* species from the Permian Period (about 299 million years ago) and *Dicroidium* species from the Triassic Period (about 250 million years ago) were widespread across much of Australia, Antarctica, South Africa, New Zealand, South America and India. This palaeobotanical evidence was also supported by the occurrence of fossil reptiles such as *Lystrosaurus*, *Cynognathus* and *Mesosaurus* which also were found in these modern locations.

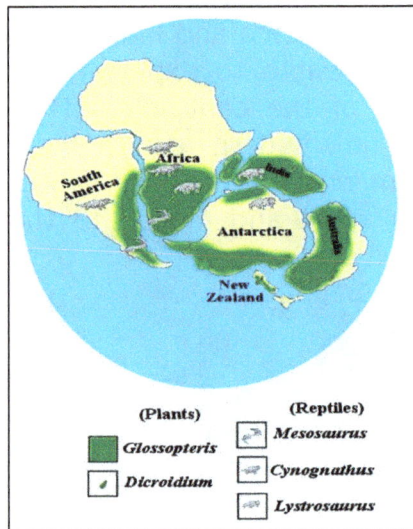

Figure 4.10: Diagram showing some of the fossil evidence for continental drift

- Matches also could be made between very old Precambrian rocks or shields which form the oldest rock units of these modern continents in the Southern Hemisphere parts of this ancient super continent. These represented the old rocks of this ancient super continent and

Figure 4.11: Diagram showing positions of ancient rock zones (Shields).

- Matching the edges of glaciation and glacial sediments which were found in these modern continents also suggested that they were once combined together.

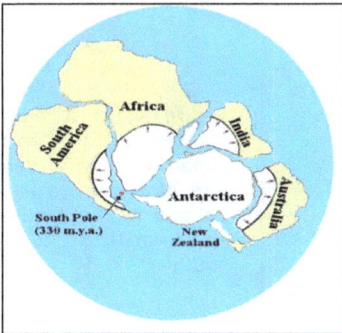

Figure 4.12: Diagram showing of glacial action about 330 million years ago.

A name was given to this ancient supercontinent which began to break apart about 175 million years ago - **Pangaea**, from the Ancient Greek Pan for all and Gaia for Mother Earth. From studies of the different parts of Pangaea, and how it broke apart piece-by-piece, it was seen to have consisted of two sections: a northern part called **Laurasia** named from the Laurentian Shield area of North America and Eurasia, and a larger southern part called **Gondwanaland**.

Surrounding Pangaea during the Late Palaeozoic and Early Mesozoic Eras, was the great **Panthalassic Ocean** from Ancient Greek for all ocean. It included the young Pacific Ocean to the west and north, and the **Tethys Sea**, named by Eduard Suess after the sister and consort of Oceanus, the ancient Greek god of the ocean, to the southeast. Following the closing of the Tethys basin and the breakup of Pangaea, the Atlantic, Arctic, and Indian Oceans were created.

Figure 4.13: A diagram showing the ancient supercontinent of Pangea and its two main parts - Lauasia in the north and Gondwanaland to the south

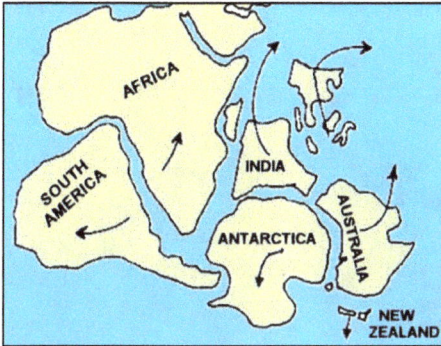

Figure 4.14: A diagram showing the possible breakup of Gondwanaland in the Late Jurassic (about 140 million years ago)

- The age of the sea floor, when measured by rock cores taken from below the ocean sediments, proved to be a lot younger than the nearby continents. This was a starling observation at a time when it was thought that there should not be any differences between the ages of the land and rocks just offshore. In some cases there were great differences in these ages, for example off the northern coast of New South Wales, Australia, the age of the old basement rocks called the Neranleigh-Fernvale Beds, have a minimum age of about 332 million years, whereas the oldest sea bed rocks just offshore in the Tasman Sea are only about 100 million years old. In other words, the nearby oceanic crust is far younger than the rocks of the nearby continent! Oceanographic surveys across many of the world's oceans showed this trend to be on a global scale.

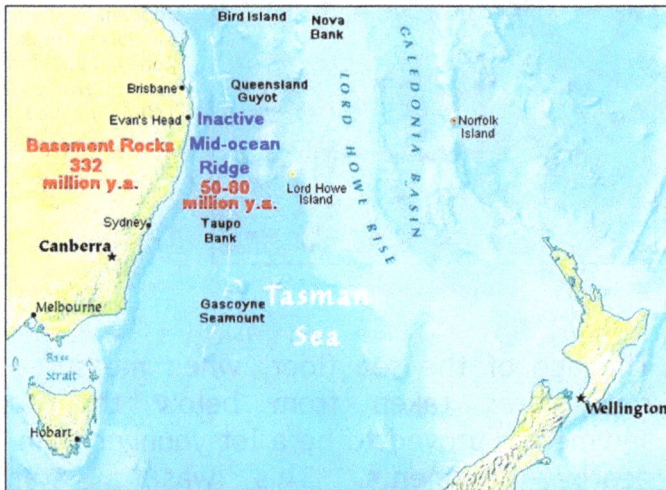

Figure 4.15: Diagram showing the relative ages of the Tasman Sea and the nearby Australian continent

- Evidence from magnetometer surveys across the oceans. Patterns of magnetic stripes seen on chart recordings of **magnetometer** surveys across the undersea mountain ranges of the oceans, termed mid-ocean ridges, showed a symmetrical pattern of magnetic intensity and direction on either side of the ridges with the youngest rocks in the centres of the ridges. As with land-based magnetometer studies, they also suggested that at times, the polarity of the Earth's magnetic field had reversed several times throughout geological history giving a **polar wandering** effect.

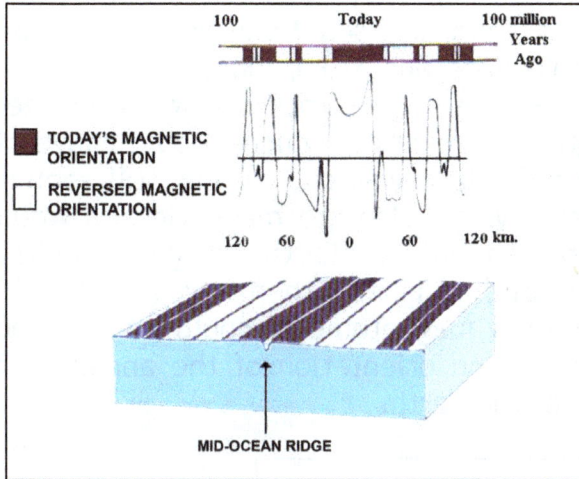

Figure 4.16: Cross-section through a mid-ocean ridge showing the magnetometer read-outs and interpretation of the magnetic stripes

Magnetometers are very sensitive instruments which use detecting coils and electromagnets which interact with the Earth's natural magnetic field and produce a movement of a pointer on a dial. The first magnetometer, invented in 1833 by **Carl Friedrich Gauss** (German: 1777-1855), used a permanent magnet and today there are various types of magnetometers including very sensitive laboratory magnetometers which can measure the magnetic field strength and direction in rock specimens collected in the field and also portable versions which can be towed behind ships, aircraft or used in satellites. Now field specimen of basalt from old lava flows could be analyzed in the laboratory for their magnetic properties. In both basalts from the sea floor going out from mid-ocean ridges and from those old flows on land, thin-sections of the rock showed tiny needles of magnetite mineral which would have formed whilst the lava was cooling but still semi-molten. These crystals also are magnetic and would have interacted with the Earth's magnetic field at that time and aligned themselves with it. The laboratory magnetometer could measure both the strength and orientation of the ancient magnetic field of the Earth.

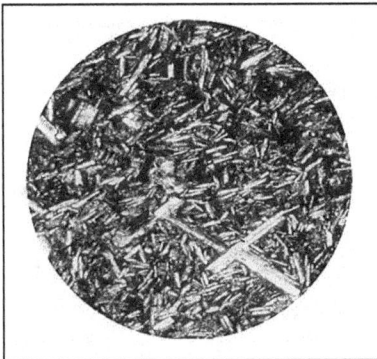

Figure 4.17: A thin section through a core sample of basalt showing the different orientations of the long, bright feldspar crystals and the smaller black magnetite crystals.

From these many observations, the idea slowly developed that the Earth's surface was not rigid but had a seafloor which was able to move and spread apart from the mid-ocean ridges forming new crust as cooled basalt lava below the oceans. This hypothesis suggested that this movement was caused by convection currents below the Earth's crust. This idea was formally proposed as the

Theory of Sea-floor Spreading in the 1960's by **Harry Hess** (American: 1906 – 1969).

On land as well as below the sea, further discoveries led to the revision of Alfred Wegener's earlier idea that the continents had also moved apart. He had put forward his **Theory of Continental Drift** in 1912, but it was widely opposed because he could not explain how such huge masses could move apart. Unfortunately, at that time, Wegener's concept of the separation of continents was not understood and many influential geologists at the time resented the views of an outsider meteorologist proposing such a hypothesis. Unfortunately, Wegener was unable to present further evidence to explain his ideas as he died in 1930 on his last expedition to Greenland. He was only fifty years old. However, mounting evidence for his theory began to emerge so that by the mid-20th Century the opinion of the scientific community had begun to change.

When a great number of these magnetic orientations were studied from a great many locations and from different geological times, it was found that the Earth's magnetic poles were in different locations than they are today, including many which were nowhere near the north and south axes. Some evidence even suggested a reversal of polarity of the Earth's magnetism. Later, this evidence would also show the directions of the wandering continents as well, but with allowances for these movements, it became apparent that the Earth's magnetic field not only wandered slightly around the north- south axis but at times it had completely reversed several times.

Furthermore, research into the nature of the sea floor, volcanoes, earthquakes and other studies conclusively showed that the surface of the Earth was still in motion. This was shown by:

- Plotting of locations of major earthquake and volcanic eruptions on a world map showed these locations as long lines or curves. This strongly suggested that the surface of the Earth may consist of large belts of volcanic and earthquake activity which enclosed places of stability. Of particular note was the grouping of oceanic trenches, volcanoes and earthquake locations around the Pacific Ocean. This has been termed the Pacific Ring of Fire. These belts also inferred that the Earth's surface down to possibly many tens or hundreds of kilometres, consisted of many moving plates, now called **tectonic plates.**

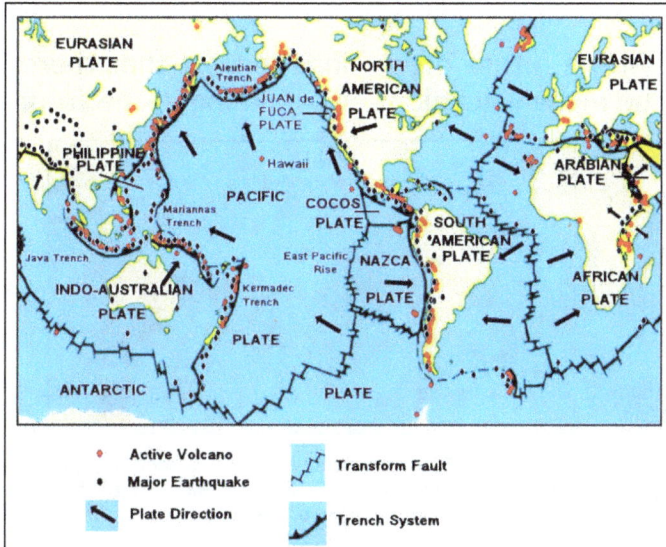

Figure 4.18: Diagram showing a simplified view of the World's larger Tectonic Plates

- The existence of hot spots at various places on the Earth's surface, often within a plate, where an excessive amount of heat was coming up from below. The **geothermal gradient** or the amount of heat at different depths, below the surface can be measured accurately. Often these hot spots formed at the ends of lines or sinuous curves which included ancient or currently active volcanoes or both. For example, the Hawaiian chain of volcanoes runs in almost a straight line from northwest to southeast where the current hot spot is located and which causes the eruption of the volcanoes on the island of Hawai'i and further out to sea where future predictions suggest that there will be more volcanic island-building in the future.

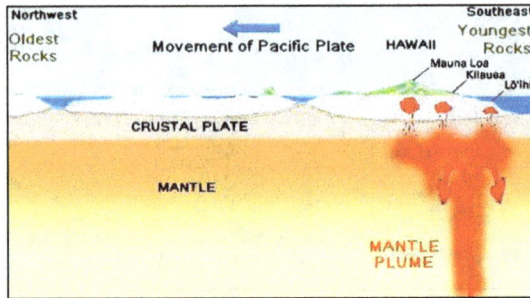

Figure 4.19: Diagram showing a cross-section of a hotspot below the Pacific Plate at Hawaii and the new submarine eruption of the Lō'ihi Seamount

- Actual measurement across a plate bounding using modern LASER **geodometers** can now be performed. These devices can very accurately measure distances such as the separation across **rift valleys** in Iceland and Africa as the plates move apart. This movement has also been accurately charted using GPS satellite systems.

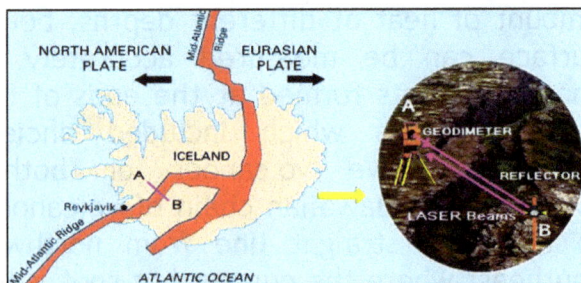

Figure 4.20: Diagram showing how a LASER geodometer can be used to measure the movement of plates across a rifting valley – here at the boundary between the two plates at Iceland

4.4 A Unifying Idea - The Theory of Plate Tectonics

It took the impetus of investigations in oceanography and geology in the second half of the Twentieth Century to dispel the old geosynclinal theory, and to show that the sea-floor formed by spreading with the continents drifting apart on large crustal plates across the Earth's surface.

By the 1980's, the Theory of Plate Tectonics, from the Latin *tectonicus* meaning to build, became the dominant theory used in explaining most of the dynamics of the Earth's **crust** – especially the large-scale motion of the continents, the formation of the oceans and the distribution and cause of volcanoes and earthquakes. In this theory, the upper part of the Earth's surface is seen as divided into two main layers called the:

- **Lithosphere** from the Greek *lithos* for rocky, and *sphaira* for sphere, is the rigid outermost shell of the Earth consisting of its the crust and upper mantle. This is broken up into tectonic plates which are composed of thinner oceanic lithosphere and thicker continental lithosphere, each topped by its own type of thin crust and the underlying:

- **Aesthenosphere** from Greek asthenes for weak and *sphaira* for sphere, is the highly viscous and ductile region of the upper mantle which lies below the Lithosphere at depths between approximately 80 and 200 km below the surface.

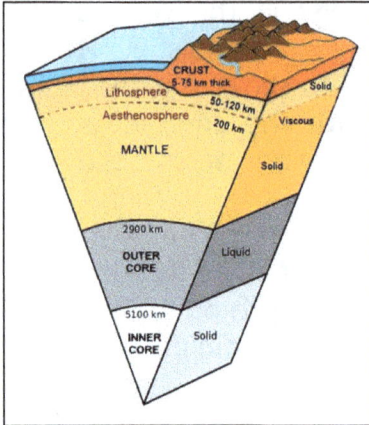

Figure 4.21: Diagram showing a cross-section through the Earth

The lithosphere, consisting of the **crust** and the solid upper **mantle,** moves on the semi-liquid and viscous aesthenosphere below. There is still some doubt as to what causes this movement. In the original theory, it was suggested that this was done by **convection cells**, an idea first proposed by **Arthur Holmes** (English: 1890-1965) in the 1930's, which resembles the way that heat is transferred through the atmosphere. Lately this convection theory has been much debated because modern computer modeling based upon 3D imagery into the Earth using earthquake waves, has not shown much support for large-scale convection cells. Some scientists have suggested that super convection **plumes**, relatively isolated up-flowing heat currents coming from the deeper mantle, actually drive the plate movements rather than large cells of convection. Another theory, that of **surge tectonics** suggests that the mantle consists of neither convection cells nor large plumes, but rather as a series of channels just below the crust allowing movement of the lithosphere above.

Whatever the mechanism, tectonic plates move around the surface of the Earth in various directions and at various speeds. These speeds may vary from about 1.0 to 8.0 cm/year according to NASA GPS data. With this movement, tectonic plates have several different types of interaction and boundaries:

- **Divergent plate boundaries** such as at mid-ocean ridges which are the places where new plate material is made from molten basaltic lava which wells up from below and spreads out horizontally to form new ocean floor rock.

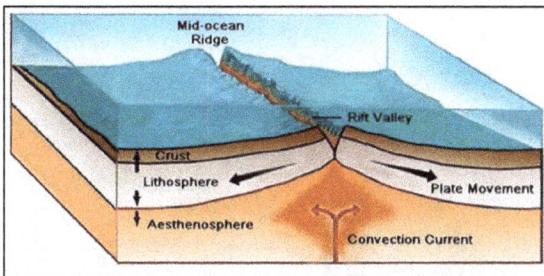

Figure 4.22: Diagram showing a mid-ocean ridge and trench system as a divergent boundary.

- **Convergent plate boundaries**, where tectonic plates meet. These occur as three types of collision:

1. Ocean-ocean convergence where two oceanic plates collide, with one plate being pushed below the other forming a **subduction zone**. As the seawater, ocean sediment and lithosphere are pushed down, they may partially melt and form andesitic magma which then rises due to its lighter density. This may form a chain of active volcanoes along the far side of the deep ocean trench caused by the subduction zone as **andesitic island arcs**, such as Japan, the Aleutian Islands and others around the Pacific Ocean rim. Because the movement down the subduction zone does not occur smoothly but in sudden bursts due to the frictional build-up of the surface of both plates, earthquakes may occur at any frictional point along the subduction zone.

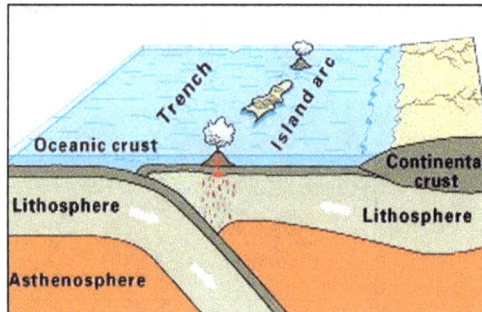

Figure 4.23: Diagram showing ocean-ocean convergence (Photo: USGS)

2. Ocean-continental convergence occurs where an oceanic plate meets a continental plate. Here, the lighter oceanic plate is subducted beneath the thicker continental plate with the re-melting of trench sediments and parts of the lithosphere to form a chain of andesitic volcanoes along the buckling edge of the continent. The Rocky Mountains and the Andes from Alaska to Antarctica are of this type.

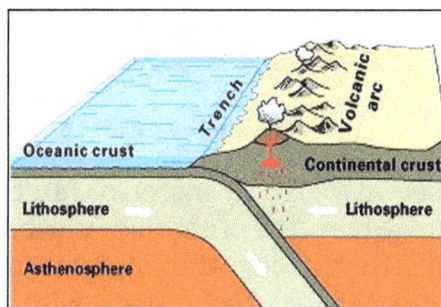

Figure 4.24: Diagram showing ocean-continent convergence (Photo: USGS)

3. Continent – continent convergence occurs when two, relatively thick continental plates collide. Here there is extensive buckling and warping of the continental crust above the deep subduction zone resulting in extensive, large area mountain-building or **orogeny**, occasional severe earthquakes, but limited volcanism. The Himalaya Mountains, the Tibetan Plateau and the adjoining Karakoram and Hindu Kush ranges are examples of this

type of collision as the southern continental plate of India rapidly moved north and collided with the Eurasian Plate.

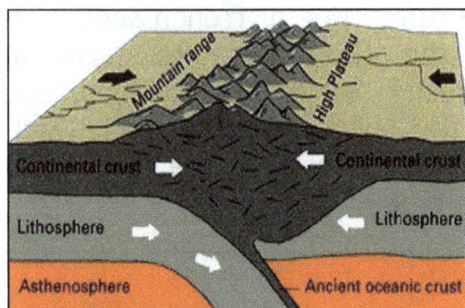

Figure 4.25: Diagram showing continent-continent convergence (Photo: USGS)

4. Conservative or **transform fault** boundaries occur when the tectonic plates slide past each other along the margins of very long fault lines in the crust with very little vertical movement. Because of the lack of vertical movement, lithospheric material is neither created nor destroyed, but a sudden movement sideways along these **faults** can occur when friction suddenly gives way. This causes major earthquake, such as those experienced along the San Andreas Fault in California and along the Transalpine Fault in New Zealand. Transform faults end abruptly and are connected on both ends to other faults, ridges, or subduction zones. Most transform faults are hidden in the deep oceans where they form a series of short

zigzags to allow the movement of rigid, flat plates over the curved surface of the Earth.

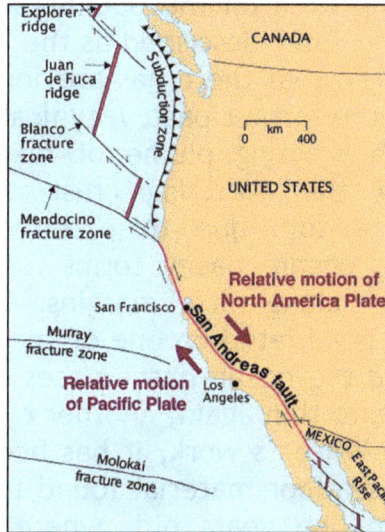

Figure 4.26: Diagram showing the transform fault boundary connecting the American and Pacific Plates (Photo:USGS)

The systems described so far have only been the most recent and therefore, the most easily studied episode of plate tectonics. The ancient super continent of Pangaea probably formed only about 300 million years ago and began to break apart about 175 million years ago. In geological time, these ages are relatively young, and there have been suggestions from palaeomagnetism and geological data that there have been other cycles of supercontinent formation and breakup in the distant past. This hypothesis has been supported by the ideas of J. **Tuzo Wilson** (Canadian: 1908-1993), who's work on the opening and closing of ocean basins, later described as the **Wilson Cycle**, also suggested that there has previous continental breakup in the distant past. A typical Wilson cycle begins with a rising plume of magma and the thinning of the overlying crust. As the crust continues to thin due to extensional tectonic forces, an ocean basin forms and sediments accumulate along its margins. Subsequently subduction is initiated on one of the ocean basin's margins and the ocean basin closes up. When the crust begins to thin again, another cycle begins. In addition to Wilson's work, it has been found that the oldest seafloor material found today dates to only 170 million years old, whereas the oldest continental crust material found today dates to at least 4 billion years old. It is reasonable to suggest therefore, that there have been other, more ancient cycles of plate tectonics in the more distant past.

Summary

1. Oceanography is the study of the sea and the sea-floor, that is, it embodies all studies concerned with the chemical and physical nature of seawater, ocean currents and seafloor topography and structure.

2. There is considerable overlapping with hydrography which is more an applied science dealing with the waters of the Earth, their mapping and study for trade and economic use.

3. Navigation became one of the first systematic studies of sea-faring people and used a variety of new instruments from stick charts of pacific, sun compasses and polarizing crystals of the Viking to more advanced devices for finding position at sea using the angle of the Sun and Stars.

4. In the Northern Hemisphere, mariners could find north by dropping a line down to the horizon from the north star (Polaris in Ursa Minor), but in the Southern Hemisphere, mariners had to find the intersection of a line drawn from the Pointers (Alpha and Beta Centaurii) to that drawn down the long axis of the Southern Cross (Crux). Dropping a line down from this point (the South Celestial Pole) locates south on the horizon.

5. Sighting devices such as Jacobs staffs, quadrants and sextants were used to finding the **parallels** of latitude (north or south of the Equator which is at zero degrees), but to find meridians of longitude was a matter of careful plotting of courses (dead reckoning) and the use a chronometer – a very accurate and seaworthy clock. The first reliable chronometer was invented by John Harrison in 1761. Meridians of longitude are measured east or west from Greenwich, London, the prime Meridian at zero degrees of longitude.

6. With the advent of the magnetic compass, mariners could plot direct courses provided they remembered that the North Magnetic Pole was a few degrees off the North Geographical Pole. The difference, or declination had to be added or subtracted depending upon its value at any specific location such that: (true north) = (magnetic north) + (+/- declination).

7. In modern times, ships used gyro compasses, which worked on the interaction between a spinning gyroscope and the Earth's rotation, rather than using the Earth's magnetism because magnetic compasses were affected by the iron of ships. Later, satellite global positioning systems (GPS) using triangulation of radio waves, became the usual navigation tool.

8.   The systematic study of the oceans gave a great tradition of sea-faring knowledge but it was not until the Challenger Expedition (1872 to 1876) of the British Royal Society that modern oceanography began to develop as a true science.

9.   From this time, a variety of instruments were used to measure and record the nature of the world's oceans and air above them. Simple mechanical devices included plankton nets, dredges and grabs, corers, thermometers and heat probes, Nansen and Niskin Bottles to sample water and bathythermographs to sample pressure and temperature. These were later supplemented by more sophisticated electronic equipment such as side-scan sonar, towed magnetometers and conductivity temperature depth arrays (CTD), as well as more direct observational tools including remote and manned submersibles and deep-sea drilling.

10. Tides are the apparent changes in the local levels of the sea and are due to the gravitational pull of the Moon and to a less extent, the Sun. When the Sun and the full moon are in the same line, high tides slightly higher than is usual occur as spring tides, and when the Sun and quarter moon are at right angles, high tides are lower than is usual and are called neap tides.

11. Tides occur twice a day (i.e. diurnal) but can also occur as two high waters and two low waters each day (i.e. semi-diurnal). These do not happen at the same time each day because the Moon takes slightly longer than 24 hours to line up again exactly with the same point on the Earth thus the timings of highs and lows are staggered throughout the course of a month, with each tide commencing approximately 24 hours and 50 minutes later than the one before it.

12. Winds are caused by the differential heating of the Earth's surface and the atmosphere above it. The equatorial region heats up more than the polar regions, and so at the equator, hot air rises leaving behind a region of lower air pressure, and at the poles the colder air will sink and accumulate giving higher air pressure. As gases move from regions of high Pressure to regions of low pressure, the polar winds flow in towards the equator until they sink at about 30^0 north and south as part of the Hadley Cell system where they form another high pressure system, the Sub-Tropical High, also called the Horse Latitudes.

13. Most of the observed atmospheric conditions including winds, storms and clouds occur within the Planetary Boundary Layer of the troposphere, the lowest of the major layers of the Earth's atmosphere.

14. The Earth's atmosphere consists of several major layers: the troposphere; the stratosphere; the mesosphere; the thermosphere; and the exosphere in order from the surface and into outer space. The ionosphere is an overlapping layer of charged particles which cause radio interference.

13. At the equator, much of the momentum of the winds has been lost and the region is a place of low pressure with warm air generally rising – this is the region known as the doldrums.

15. The World's winds are deviated from a straight line by the rotation of the Earth. This is called the Coriolis Effect, which states that moving objects on the surface of the Earth such as wind masses, are deflected to the right with respect to the direction of travel, in the northern hemisphere and to the left in the southern hemisphere.

16. Cyclones, also called typhoons or hurricanes, occur in the hottest months of late summer but other regions can experience distinct cyclone seasons, such as in northern Australia and the southern Pacific (November to April) and in the northern Atlantic (June to November). The warmer water experiences rapid evaporation with the water vapour rising in a circular motion due to the Coriolis Effect – clockwise as seen from above in the Southern Hemisphere and anti-clockwise in the Northern Hemisphere.

17. Weather details are shown on weather maps called synoptic charts which show isobars - lines joining places of equal air pressure which is measured in hectoPascals (hPa) with normal air pressure being 1022 hPa or 1013 millibars, 760 mm. or 29.8 inches of mercury, or 1 atmosphere. Low pressure (e.g. below 1022 hPa) usually means bad weather and high pressure usually means fine weather.

18. Ocean currents are caused by the prevailing wind direction, salinity and temperature differences, the Coriolis Effect of the rotating Earth, ocean-floor contours and the restrictions of landmasses.

19. Gyres are large rotating currents within the middle of the major oceans. There are five major gyres and several smaller ones.

20. There is also a sub-surface circulation deep below sea level. This is the thermohaline system, which is a great conveyor belt of deep water currents which travel at great depths around the sea floor. These are produced as warm surface currents travel into polar regions, cool (becoming more dense) and then sink as a deep current which flows back down into the major ocean basins.

21. Differences in temperature of ocean currents can cause extreme, large scale weather events including the El Niño effect, or the climatic event due to the movement of the warm pool of warm from the central Pacific across to its eastern side due to a weakening of the trade winds which brings rain and storms to the Americas and brings draught in Australia and eastern Asia. The La Niña effect in when the sea surface temperature across the equatorial eastern central Pacific will be lower than normal, and warmer in the west bring draught in the east and wetter conditions in the western Pacific.

22. The floor of the ocean is generally about 4000 metres deep and covered with fine sediments such as calcareous and siliceous ooze with some carbonate deposits. This is broken by deep trenches, such as the Marianas Trench which is over 11 km. deep, seamounts (undersea mountains) and guyots (flat-topped seamounts).

23. In warm, tropical seas where the water is clear, coral may form large barrier reefs or fringing reefs around islands. When these island sink, the fringing reefs grow as atolls.

24. In the second half of the 20th century, new methods of oceanography lead to a replacement of the old Geosynclinal Theory by that of Plate Tectonics.

25. The Theory of Plate Tectonics states that the surface layer of the Earth consists of the lithosphere which consists of several plates which move about upon the viscous aesthenosphere below. This theory has been most useful in explaining sea-floor spreading, continental drift, volcanism and earthquake locations.

Practical Tips

1. Navigation is both a science and an art. It must be learned thoroughly and it is recommended that one is trained professionally beyond the scope of this text.

2. Going to sea or on any water in a small boat is always risky. Make sure that the boat is in the best condition (motor tested, maintained) and always take backup such as oars (spare motor if an extended period is planned) and spare sails if it is fitted for sails. GPS locator or EPERB (emergency position indicating radio beacon), charts, a good compass (in addition to a GPS navigator if carried), spare water, food, fuel, wet and cold weather gear and sun-shade are essential.

3. Most maritime countries have a Coast Guard or Volunteer Coast Guard or rescue organization with which one should log the proposed voyage with dates, times and location of arrival. Remember to report in when the destination has been reached.

4. Be familiar with local boating conditions – tides, currents, sand bars, reefs and usual direction and seasons for high winds and storms. Keep and use a log book, and take advice of locals (especially fishermen). Only crazy people or their unfortunate rescuers go to sea when a storm is forecast.

5. Before departing, find out about local weather conditions for the time at sea. A small aneroid barometer is useful if one does not have a more sophisticated electronic/computer system of weather forecasting. Always watch for any approaching weather change such as an increase in cloud, or a change in the cloud type, or a weather front which is seen as a long, approaching band of cloud, especially in the direction of known storms. Don't wait! Seek shelter as soon as there is a concern remembering that most storms often approach with unexpected speed and consider the time required to return to shore.

6. Coral reefs are attractive but often hidden, so know the appropriate channel through the reef. Coral is sharp, so wear appropriate diving clothing if swimming on the reef and try not to damage the reef by walking on it. Be aware of any local dangers such as marine life (e.g. sharks, barbed fish, box jelly-fish, sea-snakes) and in some tropical regions such as the Great Barrier Reef of Australia, there are certain dangerous seasons.

Multichoice Questions

1. The following is a view of the night sky in the southern hemisphere as seen from the deck of a ship:

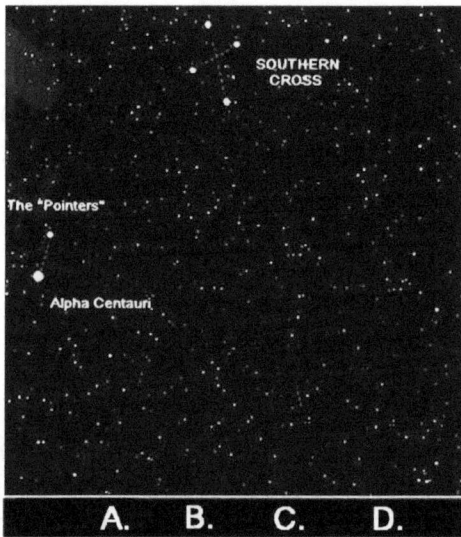

The position on the horizon that would most likely be the approximate location of south is:

A. A
B. B
C. C
D. D

2. One of the first persons to calculate the circumference of a spherical Earth was:

A. Eratosthenes of Cyrene;
B. Leif Erickson
C. Christopher Columbus
D. James Cook

3. A yacht sails from Brisbane, Australia (coordinates are: 27.47° south latitude and 153.03° east longitude) due <u>east</u> for one and a half hours. The new approximate position of the yacht would be:

A. 99.97 S and 153.03 E
B. 27.47 S and 175.33 E
C. 99.97 S and 175.33 E
D. 27.47 S and 150.53 E

4. The angular difference between geographical north and of magnetic north at a given location is called the:

A. Azimuth
B. Deviation
C. Inclination
D. Declination

5.   A method of measuring the salinity of the sea at a depth of 1000m would be to use a:

   A. Bathythermograph
   B. Piston Corer
   C. CTD Array
   D. Plankton Net

6. The following is a weather satellite photograph of a tropical cyclone near Florida, USA (north is at the top of the photograph):

From the photograph, one would expect that the wind at Miami would be coming from the:

   A. Southwest
   B. Northeast
   C. North
   D. South

7. This question refers to the following map which shows some of the Tertiary age volcanic sites in a major landmass. Numbers at the sites represent the age of the rock in millions of years from the present.

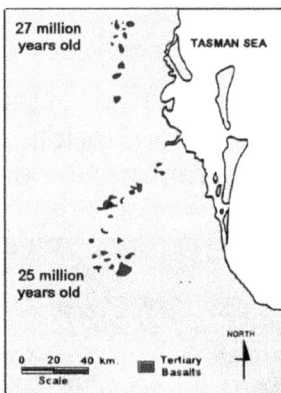

One theory for the formation of these volcanic sites considers that this part of the continent was moving over a hot spot within the mantle.

The position of these sites suggests that this continent was moving towards the:

    A. North
    B. South
    C. East
    D. West

8. The <u>best</u> statement about the theory of plate tectonics is that it:

   A. Explains everything and requires no more research
   B. There are still some problems which require further research
   C. Is based on only limited research
   D. Is only applicable for the Southern Hemisphere

9. New lithosphere is usually produced in large amounts at:-

   A. Transform fault boundaries
   B. Conservative plate boundaries
   C. Convergent plate boundaries
   D. Divergent plate boundaries

10. The following diagram shows a cross-section through the upper layers of the Earth at a subduction zone:

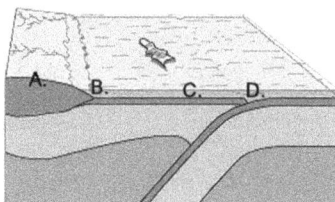

The most likely place for a volcanic eruption to occur would be:

A. A
B. B
C. C
D. D

Review and Discussion Questions

1. Oceanographers often refer to **on station** and **underway** activities when it comes to measurement and gathering samples.

(a) What do you think are the likely meaning of these two terms and
(b) List some of the instruments which may be used for <u>each</u> activity.

2. Explain how latitude is found when at sea. What are some of the difficulties that could occur in finding latitude prior to the use of GPS instruments?

3. What is the Coriolis Effect? How does it affect the world's tides?

4. Describe some of the indicators that might suggest that a bad change in the weather might occur later in the day.

5. What are the horse latitudes and the doldrums? What was their significance in early sea transportation?

6. What factors determine if a coral reef will be formed? Give some locations in the World's oceans where reefs are found.

7. Use the internet to research what are some of the present and future dangers to healthy reef development in these regions.

8. What are the main differences and similarities between atolls, seamounts and guyots, and give an explanation of how each is formed.

9. Northern Scotland and Iceland are much closer to the Arctic than Newfoundland in Canada, yet the weather in Newfoundland is much colder than these countries in northern Europe. Explain.

10. Limestone hills and mountain ranges containing marine fossils (corals, brachiopods and the like) have often found well inland and often at a considerable height above sea-level. Use the theory of plate tectonics to explain this observation.

11. Explain the differences between divergent, convergent and conservative plate boundaries, relating them to the maintenance of crustal material.

12. How does the theory of plate tectonics explain:
        (a) continental fit
        (b) sea-floor spreading
        (c) the ring of fire
        (d) the existence of *Dicroidium* fossil plant
           species in India.

13. Assuming that climate change will mean an increase in the global temperature, what are the probable effects that this will have on:

(a) the land and
(b) the oceans.

14. Evidence suggests that the breakup of Pangaea was only the most recent of several older episodes. Suggest current locations in the world which might represent new plate formation or disintegration (Hint: use the internet and look for information on the Wilson Cycle and rift valleys)

15. Use the Internet to research how modern oceanography is used by the world's maritime nations and how does this related to the use of economic resources?

Answers to the Multichoice Questions

Q1. C Q2. A Q3. B Q4. D Q5. C Q6. A Q7. A
Q8. B Q9. D Q.10 C.

Reading List

Able, K.W., et.al, (1987), Sidescan-sonar as a tool for detection of demersal fish habitats. *Fishing Bulletin*, v. 85, n. 4.

Ahrens, C. Donald, (2007). *Meteorology today: an introduction to weather, climate, and the environment. Cengage Learning. pp. 296.* ISBN 978-0-495-01162-0.

Australian Bureau of Meteorology (BOM): http://www.bom.gov.au/australia/

Bernáth, B., Farkas, A. Száz, D., Blahó, M., Egri, A., Barta, A., Åkesson, S. & Horváth, G. (2014). *How could the Viking Sun compass be used with sunstones before and after sunset? Twilight board as a new interpretation of the Uunartoq artefact fragment.* London: The Royal Society. 26 March 2014.DOI: 10.1098/rspa.2013.0787

Brooks, H. E. (2004). *Estimating the Distribution of Severe Thunderstorms and Their Environments Around the World.* International Conference on Storms. Brisbane, Queensland, Australia.

Dodds, D. (2001). *Modern Seamanship : a comprehensive ready-reference guide for all recreational boaters.* Guilford, Conn.: The Lyons Press. ISBN 9781585745289.

Fish, J.P. and H.A. Carr, (1991). *Sound Underwater Images, A Guide to the Generation and Interpretation Of Sidescan Sonar Data*. Second edition. Orleans, MA: Lower cape Publishing.

Genz, J., Aucan, J., Merrifeld, M., Finney,B., Joel,K. & Kelen, A. (2009) .Wave Navigation in the Marshall Islands. *Oceanography*, Vol. 22, No. 2, pp. 234-245.

Gnanadesikan, A., Slater, A., R. D., Swathi, P. S., and Vallis, G. K. (2005). The Energetics of Ocean Heat Transport. *Journal of Climate* 18 (14): 2604-16. Bibcode: 2005JCli...18.2604G. doi:10.1175/JCLI3436.1.

Heinemann, B. and the Open University. (1998). *Ocean Circulation*. Oxford University Press.

Howe, K. R (2006), *Vaka Moana: Voyages of the Ancestors - the Discovery And Settlement of the Pacific*, Albany, Auckland: David Bateman, pp. 92-98

Lurton, X. (2002). *An Introduction to Underwater Acoustics, Principles and Applications*. Springer in association with Praxis Publishing, pp 680. ISBN 978-3-540-78480-7.

National Geographic. (2009). *North Magnetic Pole Moving East Due to Core Flux*. December 24.

National Aeronautics and Space Administration (NASA): https://www.nasa.gov/

National Oceanic and Atmospheric Administration (NOAA): http://www.noaa.gov/

Pinet, P. R. (1996). *Invitation to Oceanography*. Eagan, MI: West Publishing Company. ISBN 978-0-314-06339-7.

Rice, A.L. (1999). *Understanding the Oceans: Marine Science in the Wake of HMS Challenger*. Routledge. pp. 27-48. ISBN 978-1-85728-705-9.

Sverdrup, K. A., Duxbury, A. C. & Duxbury, A. B. (2006). *Fundamentals of Oceanography*, McGraw-Hill, ISBN 0-07-282678-9

Urick, R.J. (1983). *Principles of Underwater Sound*. Peninsula Pub., 423 pp.

Thurman, Harold and Alan Trujillo. *Introductory Oceanography*.2004.p151-152. Prentice Hall; 5th edition (1996). ASIN: B000OIPPTO

Key Terms Index
(Page numbers in brackets)

abyssal plain (80) is the vast extent which makes up the flat sea floor.

aerosols (69) consist of droplets of liquid within a gas e.g. clouds.

aesthenosphere (104) is the viscous layer of the upper mantle just below the lithosphere upon which the plates are able to move.

anacoustic zone (59) from the Greek -*ana* meaning not and *akoustikos* meaning to hear, is a zone over 160 kilometres above the surface where the air pressure is so low that sound waves do not travel.

aneroid (43) from the Greek *a*- for without and *nēron* for water, as early barometers often used water instead of Mercury. These are compact barometers which use a small, flexible metal box which change their volume as the air pressure outside changes moving a pointer on a numbered dial.

apogee (3) when the Moon is at its farthest position from Earth.

atoll (86) is a ring-shaped reef, island often as a chain of islands formed of coral surrounding an open lagoon of water. Atolls form as the central island or seamount surrounded by reefs sinks.

barrier reefs (87) are coral reefs which form offshore and parallel to a mainland coast.

bathyscaphe (textbox, p.32) is a very deep sea submersible with some mobility, which has been used to explore the deepest ocean trenches.

bathythermographs (27) are devices dropped overboard and measure the oceans temperature and pressure at given depths.

Calcium Carbonate Compensation Depth CCD (81) is the depth in the oceans below which the rate of supply of calcite (calcium carbonate) is less than the rate of at which it dissolves so that no calcium carbonate is preserved. This depth varies at times, and it is thought that when carbon dioxide levels are higher in the atmosphere, the CCD becomes shallower.

chronometer (13) is a very accurate clock used in navigation and usually set at Greenwich Mean Time (i.e. the time at the Prime Meridian).

conductivity temperature depth array CTD (28) consists of several vertical sampling devices (often Niskin Bottles) arranged in a clustered array and dropped overboard to a pre-determined depth where the device collects water samples and measures sea temperature and electrical conductivity (a measure of the salinity).

continental rise (80) is at the end of the continental slope and marks its beginning from the flat sea floor.

continental slope (80) is the sudden incline change between the outer edge of the continental shelf and the change in slope marking the beginning of the deep ocean floor.

continental shelf (80) is the relatively flat and shallow sea floor surrounding most continents.

convection cells (33, 105) are circular currents of heat in gases (e.g. air), liquids (e.g. the oceans) and solids (e.g. the Earth's mantle) caused by differential heating from a heat source which causes the hot material to rise and then sink as it cools to start the process over again.

convergent plate boundaries (107) are those edges of tectonic plates where the plates are colliding and being pushed together e.g. subduction zones.

coriolis effect (61) is the deviation of the world's winds due to the rotation of the Earth such that winds are deflected to the right (with respect to the direction of travel) in the northern hemisphere and to the left in the southern hemisphere.

crust (26,104) is the solid upper layer of the Earth which varies in thickness from 5 km (under the oceans) to 50 km (under continents).

dead reckoning (11) is the method of ocean navigation by calculating one's position, by estimating the direction and distance travelled rather than by using landmarks or astronomical observations.

dew point (45) is the temperature below which water droplets will form.

diurnal (4) tides occur twice a day as high and low tides.

divergent plate boundaries (106) are those edges of tectonic plates which are pulling apart e.g. mid-ocean ridges.

doldrums (34) is a colloquial expression for the area of low pressure calms around the Equator called the Inter-Tropical Convergence Zone (ITCZ). is a belt around the Earth extending approximately five degrees north and south of the equator. Here, the prevailing trade winds of the northern hemisphere blow to the southwest and collide with the southern hemisphere's driving northeast trade winds

El Nino (76) is the climatic event due to the movement of the warm pool of water from the central Pacific across to its eastern side due to a weakening of the trade winds.

Environmental Lapse Rate ELR (56) is the difference in temperature between the surface and the tropopause divided by the height.

equator (11) is the imaginary circle around the Earth at zero degrees latitude where the Sun is overhead twice a year.

exosphere (59) is the upper layer of the Earth's atmosphere merging into outer space.

faults (109) occur where there is movement along cracks or joints in the Earth's crust.

Ferrel Cell (34) occurs approximately from 30° to 60° north or south of the Equator between the Hadley Cell and the Polar Cell. It is a secondary circulation feature, possibly an eddy current created by the Hadley and Polar cells.

free atmosphere (54) is the upper part of the troposphere just above the Planetary Boundary Layer.

fringing reefs (86) are coral reefs that form around an island. If the island sinks, these reefs often build up and may form an atoll.

geodometers (103) devices for measuring distance with great accuracy, usually consisting of a LASER beam sent from the device to a reflector at the other end of the distance to be measured. The beam then bounces back to the geodometer which calculated the distance from the time taken for the reflection and the velocity of the LASER (= Light Amplification by Stimulation of Emission Radiation) light.

geostrophic (56) are winds which are parallel to the isobars, the imaginary lines on the weather map joining places of equal air pressure.

geosynclines (90) are long, down-turned trough into which sediments are washed. This term has largely been superseded by other terms from plate tectonics theory.

geosynclinal hypothesis (90) an outdated theory to explain the cycle of erosion and mountain building (orogeny) in which sediments filling geosynclines caused them to sink and fold as their base re-melted and produced upward intrusions of magma.

geothermal gradient (102) is the gradual increase in geothermal heat with depth.

global positioning system GPS (20) is a satellite navigation system in which the satellites transmit microwave radio signals which are picked up by GPS receivers which then use this information from any four satellites in range to triangulate the signals and calculate the user's exact location.

gnomon (8) is a vertical post used to cast a shadow on a sun compass or sundials.

Gondwanaland (95) was the southern part of the ancient supercontinent of Pangaea.

Gulf Stream (73) is a warm surface ocean current which runs from the Caribbean, across the Atlantic Ocean and then north along the west coast of Britain and then onto Iceland. It gives a more moderate climate in these countries.

guyots (89) are flat-topped seamounts (underwater mountains, usually of volcanic origin) which have been eroded and the original island or mountain has sunk.

gyres (75) large rotating currents within the centres of the major oceans due to the main currents, influence of landmasses and the Coriolis Effect of the Earth's spin.

gyrocompass (18) is a heavily-weighted device which is rapidly spun using an electric motor such that this interaction with the Earth's rotation can be used to determine the bearing of the.

Hadley Cell (33) one of the huge convection cells of circulating air within the Troposphere (lowest layer of the atmosphere) which takes air from the equator to temperate zones. Smaller cells, the Ferrell Cell and the Polar Cell also circulate air masses.

hemipelagic sediments (84) is marine sediment that consists of fine-grained biogenic and terrigenous material and differs from pelagic sediment which is composed of primarily biogenic material containing little to no terrigenous material.

horse latitudes (33) lie between 30° and 38° north and south of the Equator and are probably named from the "dead horse" ceremony when sailors paraded a straw effigy of a horse around the deck before throwing it overboard signifying the time at which they had earned, by their extended time at sea, their advanced pay and were now free of debt.

hydrography (1) an applied science dealing with the waters of the Earth, their mapping and study for trade and economic use.

inertia (18) is the property of any mass to resist its change in position.

International Date Line IDL (12) is the meridian of longitude approximately at 180° east and west where there is a loss (going east) or gain (going west) of one day.

Inter-Tropical Convergence Zone Itcz (34) is a belt around the Earth extending approximately five degrees north and south of the equator where there is a low pressure zone of calms. Here, the prevailing trade winds of the northern hemisphere blow to the southwest and collide with the southern hemisphere's driving northeast trade winds.

ionosphere (20) is the layer of the Earth's atmosphere where atoms are ionized by solar and cosmic radiation becoming ions or electrically charged atoms or groups of atoms. It lies 75-1000 km (46-621 miles) above the Earth.

isobars (56,64) are lines drawn on a weather map (synoptic chart) joining places of equal air pressure (measured by barometers in millibars, hectopascals, centimetres or inches of mercury).

isostacy (91) is the maintenance of balance or equilibrium of the surface of the Earth between the erosion of mountains, sedimentation within basins and mountain building.

jet streams (56) are fast-flowing, meandering streams of air, usually occurring in the upper troposphere just below the tropopause.

La Nina (77) is the reverse of El Niño within the El Niño–Southern Oscillation climate pattern. In which the sea surface temperature across the equatorial eastern central Pacific Ocean will be lower than normal and warmer in the west as the trade winds and the equatorial current strengthens.

larboard (9) is the old name for the port side (left as one faces the front) of a ship.

latitude (11) is the distance north or south of the Equator, along east-west lines circling the Earth known as parallels (to the Equator).

Laurasia (95) was the northern part of the great supercontinent of Pangaea.

lithosphere (104) is the material of each tectonic plate which consists of the Earth's crust and upper part of the mantle, varying in thickness from about 5 to 150 km thick.

littoral zone (fig. 4.1, p.81) is that part of the coastline between the high and low tide marks.

longitude (11) is the position east or west around the Earth along north-south circles around the globe called meridians.

magnetic declination (16) is also called magnetic variation and is the difference between the direction to the North Magnetic Pole as shown on a compass and the direction to North Geographic Pole at a particular position on the surface of the Earth

magnetic deviation (17) is the alteration of a magnetic bearing or azimuth due to the attraction of the iron and/or the electrical interference of the ship.

magnetometer (28,97) is a device towed behind a ship or aircraft which can accurately measure the Earth's magnetic field. Portable versions are also used to measure the magnetic field during land surveys.

magnetosphere (58) is the magnetic field around the earth

Magnetic North Pole (15) or south pole, is the place of maximum magnetic intensity where the magnetic field lines appear to converge.

mantle (105) is the rocky shell of the Earth below the crust and has an average thickness of 2,886 kilometres. It mantle makes up about 84% of Earth's volume and is predominantly solid but parts near its upper surface can sometimes act as a very viscous fluid.

meridians (11) of longitude are imaginary lines on charts which are parallel to and east or west of the Prime Meridian at Greenwich near London (= zero degrees longitude up to 180 degrees east or west of Greenwich).

mesopause (58) is the boundary between the mesosphere and the thermosphere above.

mesosphere (58) is the third of the atmospheric layers from about 50 km to 100 km above the surface. Its upper boundary is called the mesopause.

meteorology (37) from the Greek *metéōros* for in the sky, it is the multidisciplinary study of the weather.

millibar mb (43) is one thousandth part of a bar which is the pressure of the air which exerts a force of 1×10^5 newtons (N) per square metre.

monsoon systems (68) are seasonal changes in the direction of the prevailing winds causing wet and dry seasons throughout much of the tropics.

nansen bottles (26) and their successors, the Niskin bottle consist of a collection tube or tubes which are dropped overboard on a cable and designed to collect water samples at a specific depth where they are shut tight by a triggered release mechanism.

neap tides (3) are slightly lower tides than usual and occur when the Sun and the Moon are at right angles to each other.

NOAA (16) are the initials for the National Oceanic and Atmospheric Administration which is an American scientific agency within the United States Department of Commerce concerned with the conditions of the oceans and the atmosphere.

noctilucent clouds NLC (58) are clouds made of frozen water vapour which are most readily visible when the Sun is just below the horizon.

obliquity (9) is a measure of the tilt of a planet such as Earth.

oceanography (1) the study of the sea and the sea-floor, that is, it embodies all studies concerned with the chemical and physical nature of seawater, ocean currents and seafloor topography and structure.

ooze (85) is very fine silt on the floor of the deep oceans. It may be calcareous (carbonate rich) or siliceous (silica rich).

orogeny (91,108) is a general term for mountain building when parts of the Earth's surface is crumpled and pushed up.

ozone layer (57) is a band of air containing a high amount of the unstable form of oxygen molecule ozone, which is a molecule having three oxygen atoms. It is above from 20 to 30 km above the surface and is an important protective layer absorbing much of the Sun's ultraviolet radiation.

Pangaea (95) was the ancient supercontinent which existed in the Late Palaeozoic and Early Mesozoic Eras consisting of Laurasia in the north and Gondwanaland in the south.

Panthalassic Ocean (95) was the great and ancient ocean which surrounded Pangaea.

parallels (113) of latitude are imaginary lines on charts which are parallel to and north or south of the equator (= zero degrees of latitude up to 90 degrees north or south).

pascal (43) is the S.I. or metric system standard of pressure, and for convenience, the hectopascal is often taken as being equal to the millibar.

pelagic sediments (85) are deep marine sediments composed of primarily biogenic material and containing little to no terrigenous material.

pelorus (19) is a device used to take sightings (bearings) from the ship to distant objects. It is usually mounted on top of and aligned with the compass (magnetic or gyro).

perigee (3) is when the Moon is at its closest position to the Earth.

photoionization (58) is the breakup of molecules by light energy.

Planetary Boundary Layer PBL (54) is the lowest part of the troposphere just above the Earth's surface.

plumes (106) in the Earth's mantle are relatively isolated up-flowing heat currents coming from the deeper mantle.

Polar Cell (34) the air masses at the $60°$ north or south of the Equator are still relatively warm and rise to the tropopause. Here they move towards the poles where it cools and descends, creating a cold, dry high-pressure area at the polar surface.

polar wandering (15,97) occurs such that the Earth's magnetic poles have moved over time including complete reversals of polarity from north to south and vice versa.

port (9) is the left hand side of a ship.

precession (18) is the slow movement of the axis of a spinning body around another axis due to a force, such as gravity, acting to change the direction of the first axis.

Prime Meridian (12) is the starting point or zero degrees of longitude which runs through Greenwich near London.

relative humidity (45) is the amount of water vapour which the air contains as a percentage of the maximum amount it can hold at the same temperature. This is measured by various hygrometers.

ridges (65) are elongated areas of high pressure often between two troughs and usually bringing finer weather.

rift valleys (103) are long valleys with parallel and steep sides of tensional faults formed by the pulling-apart of a continental plates.

semi-diurnal (4) where there are two high and two low tides per day.

side-scan sonar (27) SONAR is an acronym for **SO**und **N**avigation **A**nd **R**anging and a side-scan sonar device can be towed behind a ship. This device sends out pulses of sound waves at certain frequencies which bounce off the sea-floor and other reflective objects and return to a sensor which then interprets the reflections on a computer screen. The sound waves are sent out from the ship sideways and reflect off the sea floor objects.

sólarsteinn (8) is Icelandic for Sunstone and are crystals of clear Iceland spar, a variety of calcite which would show the position of the Sun by the polarizing effect of the crystals even after sunset.

sonar (27) originally an acronym for **so**und **n**avigation **and** **r**anging, is a technique that uses transmitted sound waves underwater to detect objects on or under the surface of the water.

South Celestial Pole (7) is the imaginary spot in the sky of the southern hemisphere around which all the stars seem to rotate because of the movement of the Earth. It is immediately above south on the horizon.

Southern Oscillation (77) is a to-and-fro movement of atmospheric pressure within the tropics which produces alternations of El Niño- La Niña events.

Southern Oscillation Index SOI (77) is calculated using the air pressure differences between those at Tahiti and Darwin.

spring tides (2) are slightly higher tides than usual and occur when the Sun and the full Moon are in the same line.

starboard (9) is the right hand side of a ship.

storm surges (61) occur when the high winds push the surface of the sea outwards so that when they hit coastlines they often cause localized flooding.

stratopause (58) is the boundary of the stratosphere with the mesosphere above.

stratosphere (57) is the second layer of the atmosphere with its upper boundary at the stratopause just below the mesosphere.

subduction zone (107) is a convergent (coming together) plate boundary where one plate is forced below another.

surge tectonics (106) is a theory of heat transfer in the Earth's mantle which suggests that the mantle consists of neither convection cells nor large plumes, but rather as a series of channels just below the crust allowing movement of the lithosphere above.

synoptic charts (63) are weather maps showing pressures as isobars.

tectonic plates (101) are the large sections of the Earth's upper crust which form the surface and are in constant motion.

Tethys Sea (95) was an ancient sea formed within a major indentation between Laurasia and Gondwanaland.

Theory of Continental Drift (100) an early idea based on the probable fit of continents suggesting that the present continents were once together but spread apart.

Theory Of Plate Tectonics (119) explains the dynamics of the Earth's surface as being due to the movement of a large number of tectonic plates which are produced at Mid-ocean ridges and destroyed by sinking below or colliding with other plates.

Theory of Sea-Floor Spreading (100) was an idea proposed by Harry Hess in 1960 based on magnetic surveys of the sea-floor, which suggested that the sea-floor was made by upwelling of magma creating new rock (basalt) at mid-ocean ridges.

thermistors (41) are electrical resistances which drop their resistance to current when heated and so can be used in solid state devices for measuring temperature.

thermohaline system (75) is the sub-surface circulation between ocean currents deep below sea level and those on the surface.

thermopause (59) is the upper edge of the thermosphere separating it from the exosphere.

thermosphere (58) is the fourth layer of the atmosphere with its upper boundary as the thermopause.

topography (1) is the surface of any land above or below the sea.

tornadoes (67) is a violently rotating column of air that spirals upwards while in contact with both the surface of the Earth and a cumulonimbus cloud as a narrow area of low pressure.

transform fault (109) is a large break or fracture within the lithosphere which enables the rigid plates to move around the curved surface of the Earth. These are strike-slip faults which move sideways with little or no vertical displacement.

triangulation (20) is the method of taking at least three sightings (or bearings) from the observer or location - the directions then being drawn back on a chart or referenced to the GPS satellite to obtain an accurate position

tropical cyclones (60) also called typoons and hurricanes, are all very intense, rotating regions of low air pressure bringing very high winds and rain to places in and near the tropics. In the northern hemisphere they rotate anti-clockwise and in the southern hemisphere they rotate clockwise.

tropopause (56) is the boundary between the troposphere and the overlying stratosphere.

troposphere (33,53) is the first layer of the atmosphere with its upper boundary at the tropopause at about 9 km (at the poles) to 17 km(at the Equator) above sea level.

troughs (65) are long, extended low pressure regions which are often associated with ridges which are their high pressure counterpart. They tend to bring stormier weather.

turbidity current (80) is a sudden flow of sediment and water off the continental shelf and often down submarine canyons in it. The sediment may also be accompanied by fragments of the shelf as well. They deposit poorly-sorted sediments called turbidites.

weather front (66) is a boundary separating two masses of air of different densities, and is the principal cause of meteorological phenomena outside the tropics. Warm and cold fronts are often seen as a long line of approaching cloud.

Wilson Cycle (111) attempts to explain the opening and closing of ocean basins beginning with a rising plume of magma and the thinning of the overlying crust forming an ocean basin with sediments accumulating along its margins. Subsequently subduction is initiated on one of the ocean basin's margins and the ocean basin closes up. When the crust begins to thin again, another cycle begins.

www.ingramcontent.com/pod-product-compliance
Lightning Source LLC
Chambersburg PA
CBHW050729030426
42336CB00012B/1479